THE QUEST OF THE SCHOONER ARGUS

BOOKS BY ALAN VILLIERS

The Quest of the Schooner Argus

The Set of the Sails

Sons of Sinbad

Cruise of the Conrad

Whalers of the Midnight Sun

Grain Race

By Way of Cape Horn

Falmouth for Orders

The Coral Sea

Books for Younger Readers

Joey Goes to Sea

Stormalong

The Arctic Doryman

ALAN VILLIERS

THE QUEST OF THE SCHOONER ARGUS

A VOYAGE TO THE BANKS AND GREENLAND

ILLUSTRATED WITH THE AUTHOR'S PHOTOGRAPHS

CHARLES SCRIBNER'S SONS
NEW YORK
1951

COPYRIGHT, 1951, BY
ALAN VILLIERS

Printed in the United States of
America

*All rights reserved. No part of this
book may be reproduced in any
form without the permission of
Charles Scribner's Sons*

A

TO
DR. PEDRO THEOTONIO PEREIRA

CONTENTS

Preface		11
Prologue		15
Chapter 1	THE FLEET ASSEMBLES	19
Chapter 2	THE BACKGROUND	33
Chapter 3	ACROSS THE NORTH ATLANTIC	51
Chapter 4	INTERLUDE AT ST. JOHN'S	69
Chapter 5	ON THE GRAND BANKS	84
Chapter 6	BELLS IN THE FOG	99
Chapter 7	THE PROBLEM OF BAIT	119
Chapter 8	INTO NORTH SYDNEY	138
Chapter 9	IN THE ICE	153
Chapter 10	THE GREENLAND CAMPAIGN—JUNE	167
Chapter 11	FIRST FISHER OF PORTUGAL	187
Chapter 12	THE CAPTAINS FROM ILHAVO	205
Chapter 13	THE GREENLAND CAMPAIGN—JULY	224
Chapter 14	CRUISE IN THE *Gil Eanes*	241
Chapter 15	THE GRAND BANKS AGAIN?	263
Chapter 16	END OF THE CAMPAIGN	282
Chapter 17	THE VOYAGE HOME	297
Appendices A	THE *Argus* AND HER VOYAGES	312
B	THE PORTUGUESE GRAND BANKS FLEET	320
C	SOME ECONOMIC NOTES	328
Index		339

ILLUSTRATIONS

Frontispiece   THE ARCTIC DORYMAN

Between pages 32 and 33
>   SHE HAD A GLOUCESTER BOW
>   IN DRY DOCK HER GOOD LINES SHOWED
>   LOADING SALT
>   THE FLEET ASSEMBLES IN THE TAGUS
>   THE NESTS OF DORIES
>   BEACH FISHING TYPE, CAPARICA
>   THE LAST BARQUENTINE
>   AN OLD-TIMER ON THE WAY TO THE BANKS

Between pages 64 and 65
>   THE MAIN DECK WAS FULL OF DORIES
>   DORYMEN PREPARING THEIR LONG-LINES
>   CAPTAIN ADOLFO USING THE MICROPHONE
>   A DECKBOY IN THE HOLD SHOVELLING SALT
>   DOWN IN THE RANCHO

Between pages 96 and 97
>   WAITING FOR BAIT AT ST. JOHN'S
>   A MEAL IN THE *Argus*, AT ST. JOHN'S
>   DORYMAN WITH "GRIPPERS" TO PROTECT HIS HANDS
>   THE DORY IS TO THE DORYMAN . . .

Between pages 128 and 129
>   THE FISH WERE WASHED IN VATS OF RUNNING WATER
>   FISH CLEANERS AT WORK
>   FULL POUNDS—A GOOD DAY'S FISHING
>   DORIES BACK, GOOD WEATHER
>   THE MATE SOUNDING THE FOG BELL
>   THE HAPPY DORYMAN
>   SOMETIMES DORYMEN HAD TO SWIM FOR THEIR LIVES
>   AN AZOREAN VETERAN
>   HE HAD MADE 35 VOYAGES
>   THE BROTHERS SALVADOR AND ESTRELA MARTINS

ILLUSTRATIONS

Between pages 160 and 161

  César de Medeiros
  The Trolley-pusher
  The Excellent Chief Cook with His Pet Hen
  Deckboy with a Big Cod
  The *Argus* Lay Quietly
  The *Avis*, Sailing for North Sydney
  Bound for Greenland

Between pages 192 and 193

  In the Straits of Belle Isle
  Away Went the Dories Once More
  A Dory Swamped
  Full Dory
  The First Fisher's Dory, Full Almost Every Day

Between pages 224 and 225

  João de Oliveira Was Second Fisher of the *Argus*
  The First Fisher—Francisco Emilio Battista
  Captain of the Schooner *Santa Isabel*
  Captain Silvio Ramalheira
  The Afterguard of the *Argus*
  She Was the Barquentine *Gazela*
  A Big Old Fellow from the Arctic Depths
  The Second Mate
  The Boatswain and His Son

Between pages 256 and 257

  Antonio Rodrigues Chalão
  The *Gil Eanes*
  The Endless Job—Fixing the Long-Line
  The *Elisabeth* Sailing From Greenland
  The *Gazela* Got Under Way for Lisbon
  The Captains Enjoy a Cheerful "Gam"
  Young Dorymen

Between pages 288 and 289

  Down Davis Straits
  The End of the Voyage
  The Dorymen Went Back to Coastal Fishing

PREFACE

ONE day in January, 1949, when I chanced to be in Washington, D.C., the Portuguese Ambassador, His Excellency Dr. Pedro Theotonio Pereira, suggested that I might find it interesting to sail with the Portuguese codfishing fleet to the Grand Banks off Newfoundland, and to the Greenland fishing grounds in Davis Straits. There were, he said, still at least thirty sailing-ships taking part in this historic fishery, from Lisbon, Oporto, Aveiro, Figueira da Foz, and Viana do Castelo. Wherever there were thirty sailing-ships operating was a good place for me. I could not go with the fleet in 1949 as I had undertaken then to sail the Marine Society's school-ship *Warspite* for the Outward Bound Sea School, in North Wales. But by early March, 1950, I could go. I went, and was most grateful to the Ambassador for providing the opportunity, and for seeing that I was enabled to make full use of it. He arranged a berth for me in the excellent schooner *Argus,* and passed me to the good care of the officials of the Guild of the Codfishing Shipowners (Gremio dos Armadores de Navios da Pesca do Bacalhau) and the Chief of the Services for Assistance to the Ships at Sea.

To the Ambassador, the owners of the *Argus* (the

Parceria Geral de Pescarias, of Lisbon), and especially Senhor Vasco Bensaude, the senior partner; and to Commander Henrique dos Santos Tenreiro, the energetic Government Delegate to the Guild of the Codfish Shipowners and power behind the numerous organisations for the improvement of fishing and fishermen's conditions generally in Portugal, I owe a considerable debt. Commander Tenreiro ensured that I was able to see everything I wished to see, and to learn everything I wished to learn. Both were extensive. I am grateful, too, to Senhor Eduardo van Zeller, who took the Ambassador and myself on a grand run through the fishing ports of the Algarve; to Dr. Joaquim Bensaude, father of Vasco Bensaude and, at 92 years of age, still one of the greatest living authorities on the earlier Portuguese voyages; and to James Vinden, of the British Council at Lisbon.

At sea and in Portugal, Commander Americo Angelo Tavares de Almeida, the Chief of the Services for Assistance to the Ships at Sea and Captain of the ports of the Grand Banks and Davis Straits for the Portuguese fishing vessels, was friend, host, and invaluable informant. A former submarine commander in the Portuguese Navy (in which he is still an active officer), Commander Tavares de Almeida has been voyaging to the Banks and Greenland grounds for six months of every year since 1943. His eight long voyages have given him a sound general knowledge of the fisheries which is probably unrivalled. I spent some pleasant weeks as his guest in the hospital and assistance-ship *Gil Eanes*, which flies his flag, and tends the needs, bodily, maritime, and spiritual, of the 3000 Portuguese fishermen who toil annually in those parts.

PREFACE

I am grateful also to the pleasant Master of the schooner *Argus*, Captain Adolfo Simões Paião, Jr., of Ilhavo, to whose thirty-first voyage to the Banks and Greenland I must have added complications and duties which, though strange, he accepted cheerfully. He was a good friend, a fine host, a first-rate shipmate, and a mine of information. His brother captains—particularly his blood brother Francisco da Silva Paião in the schooner *Creoula*, Silvio Ramalheira in the *Elisabeth*, João Pereira Ramalheira (called Vitorino) in the *Gil Eanes*, and José Teiga Gonçalves Leite in the barquentine *Gazela*—were always friendly and helpful. So also were their officers and crews, especially the two mates of the *Argus*, Senhores João Fernandes Matias and José Luiz Nunes de Oliveira, and the engineers in that vessel, Senhores César Eduardo Mauricio and Manuel da Maia Rocha.

At Ponta Delgada in the Azores, St. John's in Newfoundland and North Sydney, Nova Scotia, I was assisted in my quest for the Grand Banks story by many persons, particularly by Senhor Roberto Arruda at Ponta Delgada; by Senhor João Morrais, the Portuguese Consul at St. John's, Mr. R. Gushue, C.B.E., Chairman of the Newfoundland Fisheries Board and also of the Fisheries Production Committee of the Combined Food Board for the United Nations, Mr. J. W. Allen, O.B.E. of the Furness Withy Line at St. John's, Dr. Wilfred Templeman, O.B.E., Director of the Government Fisheries Laboratory in Newfoundland; and, at North Sydney, Mr. M. Benac, for many years agent for the French and Portuguese Grand Bankers visiting that port. To all of these, and to the many other friends made on the voyage, I express my grateful thanks.

For their advice and assistance in assuring the success of the photographic side of the voyage, I thank my friends of the National Geographic Society at Washington, D.C., in whose rooms the project was first discussed. I would like particularly to thank Mr. Melville Bell Grosvenor, of the Board of Trustees.

The brief quotations from Camoens' "Lusiads" are from the translation by Sir Richard Fanshawe, which was published in 1655 and re-issued by the Harvard University Press in 1940. Nowhere does the Portuguese spirit find better expression in literature than in the Lusiads.

There remains one obligation, and that a great one. That is to offer special thanks to my friend Colonel F. C. C. Egerton, of Sudbury in Suffolk, whose special knowledge of things Portuguese has been of considerable assistance. His enthusiastic admiration for the mariners of Portugal acted as a special inducement to me to make the voyage, and at all times, his advice and help have been of great assistance in the preparation of this book.

ALAN VILLIERS

Leafield, Oxford.
January, 1951.

PROLOGUE

> They now went sayling in the Ocean vast,
> Parting the snarling Waves with crooked Bills:
> The whispring Zephyre breath'd a gentle Blast,
> Which stealingly the spreading Canvas fills:
> With a white foam the Seas were overcast,
> The dancing Vessels cutting with their Keels,
> The waters of the Consecrated Deep . . .

THE schooners came suddenly from the Atlantic mist, and the sun shone upon them in the clear patch where we were, in a Finnish sailing-ship bound homewards from Australia after a passage of the Horn. There were three schooners, close together, and I thought at first they must be big yachts upon some ocean race from Europe towards America. They were three-masters of about three hundred tons. Their lines were graceful, and their appearance in the distance, deceivingly yacht-like. Yet I could soon see that they were too deeply laden to be yachts. As they came closer, crossing our track, I saw that they were full of small boats, painted red and brown, stowed on deck in tiers that were six boats high. Their decks were full of swarthy men, and the rigging on both sides was hung with barrel buoys; and though their main decks were crowded with these small open boats, there were no lifeboats, and no davits. I saw that clearly, for, one astern of the other in a lovely line which moved effortlessly and with grace across the sea, they passed very close by our full-rigged ship, close enough to read the names—*Maria da Gloria, Neptuno II, Argus.* The swarthy men waved cheerfully to us, some of them turning from their work at the main pumps. From the *Maria da Gloria,* a stream of clear water was gushing from the scupper-holes along her lee side, back into the sea.

In a few moments the mist swallowed them again, but not before the last of them had run up a Portuguese ensign and exchanged salutes with our blue-crossed white flag of Finland. What kind of ships were these, I wondered, that sailed westbound before the east winds of Spring, across the North Atlantic, that filled their decks with boats and men and their holds with I knew not what, and were bound I knew not where?

"They're Bankers," our Frenchman said. "Grand Bankers. Fishermen that work the Banks of Newfoundland. They anchor on the Banks and they fish with those dories, and they stay there until they are full of fish, or founder."

Our Frenchman, I knew, was from the fishing port of Fécamp. Fécamp and St. Malo sent many three-masters to the Banks of Newfoundland: he had grown up with them.

"Dories?" I asked. "Dories? What are they?"

"Flat-bottomed row-boats," he said, still gazing where the three schooners had disappeared. "You saw them stacked in their nests on deck."

"And they fish in them on the open sea, here in the North Atlantic?"

PROLOGUE

"That's what they do. Those schooners are full of salt, now. When they're on the Banks, they send those dories over the side at four o'clock sharp every morning when the weather is at all possible, and the men fish all day with lines. Then they work all night cleaning the fish, and salting them down in the hold, and catching bait—if they're lucky—for the next day. When they're full they come home."

"Sounds like a tough life to me," I said. We knew something about the North Atlantic, and a little—very little—about facing it in boats. A few weeks earlier, we'd been out in one of our lifeboats fishing a shipmate out of the sea, and narrowly escaped being lost ourselves. Compared with a dory, our ship's lifeboats were ocean-going vessels, and our full-rigger was two thousand tons.

"A tough life, you say?" The Frenchman looked fiercely at us. "A dog's life, that's what it is! My God, there is no harder life upon the sea! All fishing is tough, but *that's* the toughest, hardest way to make a living that I know. Those fellows will be lucky to be back home six months from now. Aye, and some of 'em won't be coming. I warn you, shipmates, things are tough all over Europe now, but don't ever ship in one of them! Those Portuguese use one-man dories. Keep out of them!"

I stared off into the mist again, lightening now, and I could again see dimly the shapes of the three lovely schooners, ploughing their quiet ways westwards in the sea. I could see them dimly there, and I wondered what a one-man dory was, and decided upon the advice of Pierre Berthoud that I would take no active steps to find out.

THE QUEST OF THE SCHOONER ARGUS

If Pierre Berthoud, able seaman in the ship *Grace Harwar* in the year of grace 1929, thought fit to warn his shipmates against such things, they must be hard indeed. For we were then four months and more at sea, on the way from South Australia, and it was some time since we had had a square meal. Our ship was leaking, and we knew what it was to pump. We had lost men from our pitifully small crew, and were then so reduced that we could no longer muster watches, but worked the ship as best we could. One of our officers had been driven mad. There was scurvy aboard. The ship was foul with barnacles and long grass, and the Channel still a thousand miles away. The *Grace Harwar* was tough, herself.

But one-man dories, and Banking schooners—aye, steer clear of them!

CHAPTER ONE

THE FLEET ASSEMBLES

> Now in the famous Port of Lisbon Town
> (Where golden Tagus mingles his sweet Flood
> With the Salt Ocean, and his Sands doth drown),
> With noble longings, and transported mood,
> The Ships lye ready. . . .

MY SHIP would be the schooner *Argus*, the Ambassador had written, adding that she was a fine four-master built within the past ten years. She would sail from Lisbon for the Grand Banks and Greenland early in April; with luck, she should be ready to sail home again by the end of August, full of fish. If I did not wish to make the full voyage, I could leave at almost any time and take passage in the assistance-ship *Gil Eanes* to St. John's, and fly home from Gander, in Newfoundland. The *Gil Eanes* would be with the fleet all summer.

It was a bright March morning, and the *Argus* was lying with a group of schooners off Belem, at the anchorage by the tower whence so many great Portuguese voyagers had departed. She was a beauty, a graceful steel four-master of glorious sheer and lovely lines; tall masted, staunch of hull and stout of rig, fleet of bow and with a run on her like an ocean racer. That lovely white schooner a cod-hunter? It seemed hardly credible. She lay quietly with a round dozen of her kind, three and four-masters, some wood, some steel. The green hills of Lisbon made a fine background for the fleet, sitting lightly upon the broad waters of the golden Tagus that bright morning, as the liner *Andes* brought me in from England.

I saw the *Argus* in dry dock a day or two later, being cleaned and painted for the coming voyage. Her strength and grace of line showed well in the dock, and I liked the long sweep of her main deck. She looked fast, seaworthy, and able, and I knew she would have to be all these things to survive the sort of voyage she would make, and had been making. The Portuguese Banks schooners now fished in Davis Straits, well within the Arctic Circle, since the cod had thinned on the banks off Newfoundland and their every known haunt was also the haunt of trawlers. In 1950, a codfishing campaign meant a voyage first to the Grand Banks and then to Davis Straits, fishing there so long as the Arctic conditions would permit. If not full then, the ship would fish once more off Newfoundland in the stormy months of autumn. Six or seven months in any ship under these conditions would be no joke. The new *Argus* was about 700 tons, which was not a large ship for the Greenland grounds, nor for anywhere in the savage North Atlantic.

Some thin men, lithe, clean-limbed, and brown, were bending heavy cotton-canvas sails. They were doing their work well, as it required to be done.

Daily during March more fishing schooners arrived

THE FLEET ASSEMBLES

at Lisbon, assembling to fit out for the coming campaign and to be present at the blessing ceremonies. There were thirty-two sailing-ships still taking part in the hard trade—thirty-one schooners, all three and four-masters, and a last lone barquentine. This was the last working sailing fleet from all Europe: these were the last plain, honest sailing-ships, wresting a living from the deep sea, going about their work unsubsidised, fishing in the traditional manner with lines and hooks from the small rowing-boats called dories. There were nests of dories on the *Argus'* decks—ten of them, each with six dories. From the vantage-point of the dry docks' stone balustrade, I could see that the dories, which looked about fourteen feet long, were stripped down and fitted into one another, like a child's hollow blocks. How any of these sixty dories could hope to survive in an Atlantic sea I did not then know. But I was going to find out before very long. It seemed strange that men could be found, in a so-called atomic age, to venture their lives in craft so seemingly frail and undersized. There was no special flotation gear in those dories. They were without buoyancy tanks of any sort: such refinements could not be fitted in them, or they could not be nested. They were plain flat-bottomed boats built up of planks, with bottoms of three planks and sides of four, nailed together. They had not even a fitted thwart, or a center-board, or pintles to hang a rudder. Whatever might or might not happen in the coming voyage, I was certainly going to learn how some very skilful and courageous seamen operated. These schooners were large ships to fill by hand, with hooks and lines and little rowing-boats.

I watched the *Argus* with keen interest, and visited

her daily at her berth after she had left the dry dock, while she took in salt, and fuel oil, and stores. Daily she sank more deeply into the water. Little Lisbon sailing barges with curiously back-raked single masts brought her salt and stores, vast quantities of both, and barefooted men, baskets on their heads, pitched the salt into her big hold. Spare anchors, cables, canvas, cordage, paint, oil, bales of fishing line and cases of best steel hooks, planking for making dory thwarts, poles for masts and booms, bundles of oars, bales of rubber gloves and line-grippers, cases of knives, sea-boots, tobacco, barrels of salt meat, flour, beans, olive oil, olives, pigs' feet, vinegar, red wine, brandy—all these things she loaded, and more. Day after day, the stores poured aboard. Fitting out was a big job for the skeleton crew, for the main body of fishermen had not then joined. They would come when the ship was ready, a day or two before she sailed, just in time for the blessing.

One day the *Argus* slipped into the windy Tagus and was swung for compass adjustment, a matter of vital importance for any Arctic fisherman; with her were her steel sisters, *Creoula* and *Santa Maria Manuela*, both as taut and trim and well laden as she was. A fresh wind blew that morning, and the trio of lovely schooners danced lightly in the Tagus as their bows were steadied upon compass point after compass point with hair's breadth precision, while the appropriate hoists of bunting warned other ships to give them a berth, and the sun shone upon them, and their white sides glistened.

Day after day they grew slowly ready for the sea, and the Lisbon streets knew again the annual visit of the check-shirted Arctic fishermen in their distinctive clothes—always a colourful check shirt, the more brightly checked the better, and dark striped trousers (sometimes these were checked, too), canvas shoes, or perhaps rubber thigh boots turned down at the knee, and cloth cap, or a sou'wester, or a black stocking cap such as the men of Nazaré wear throughout their lives. This was the rig. Some had coats, which were always striped and of dark material. With their striped trousers and dark coats some of them looked, at a quick glance, more like bankers than the Grand Bankers they were, but a second glance would dispel that illusion. Great breadth of shoulder, strong-jawed countenances fierce of eye with the look of men unaccustomed to city stress, rolling carefree gait, and a straight look for any man—these things distinguished them, and the strange dress of their womenfolk. For a week or two, the distinctive costumes of women from small places in North and South Portugal were to be seen about the Lisbon streets in the evenings. By day the women lived aboard their husbands' ships, with their men and often with their children too, for this was a privilege allowed good men condemned to spend the half of every year bereft of their families. The Arctic fishermen and their wives stood out ashore, just as the ships they served differed from all others in the river.

Meantime in fifty hamlets and villages all round the coast of Portugal, dutiful wives, fond mothers and sisters were busy putting the finishing touches to their menfolk's fishing clothes. The bright check shirts and the even more brightly checked underwear were not factory products. Such things were sewn in the fishermen's homes with material bought from their own co-operative store in the Fishermen's Institutes. In a thousand homes, fond women sewed diligently upon garments they were destined not to see again for far too long. In the white, flower bedecked homes of the fishermen's settlement at Nazaré, where the little boys shifted from baby clothes to the garb of fishermen—check shirts, baggy trousers, stocking caps just like their fathers; in a hundred homes of distant Fuzeta, down in the lovely Algarve, that almond-laden sunlit province of southern Portugal, which in the days when there were kings in Portugal always ranked as a distinct Kingdom; in the great fishing ports of Lisbon, Oporto, Aveiro, Figueira da Foz and Viana do Castelo; in the captains' homes of Ilhavo, a village by Aveiro whence come more Banking captains than from anywhere else on earth; in the hamlets of Furadouro, Murtosa, S. Martinho do Porto—in all these and many more, women sat working. A man must have warm kit for the Arctic fishery. The owners provided sea boots and sou'westers and some pairs of woollen gloves, but a man who has no chance of getting laundry done needs a good stock of shirts, socks, and underwear. Everything that could possibly be made from material of bright checks was made as bright as it could be, even snug head-coverings with ear muffs to go beneath sou'westers, and long underwear. Fishermen's wives, officers' wives, boatswains' wives, deckboys' mothers, cooks' wives—for them all, the month of March was stitching time, the fitting-out time for ships and men.

A hundred boys, graduates of the Fishermen's School at Lisbon, were drafted to the fleet, and smaller boys from the training frigate *D. Fernando y Gloria* and from all the primary fishing schools round the coast and in Madeira and the Azores moved in to take their places. Freshly appointed officers, the ink scarce dry on

Map of Portuguese Coast, Showing Fishing Ports and Centers

25

their certificates, daily reported to their ships, where they were put to work at once on the twelve or fifteen-hour day that seems inseparable from fishing. As the beginning of the last week in March approached, the pace increased. More and more schooners and small motor-ships kept arriving—the graceful fleet from Viana do Castelo, which included the two 70-dory motor-ships *São Ruy* and *Santa Maria Madalena,* built experimentally as a Portuguese answer to the threat of trawlers in 1939, and a great success, combining the best of the old with the best of the new; the beautiful schooner *Hortense,* which with the *Santa Isabel* was perhaps the loveliest of them all; four-masters from the famous yard of the Monicas at Gafanha by Aveiro, which had a clearly discernible trace of the caravel in their perfect hulls; and with all these, an old-timer or two, still staunch and sound for yet another voyage.

Prominent among these old-timers was the lone barquentine, the pretty little *Gazela.* She was a vessel with a clipper bow and a lovely run to her long counter, and a carved figurehead at the prow. In 1950, she was making her fiftieth Transatlantic fishing voyage. She had been fishing the Banks since 1900 and Greenland since 1931: for the first seventeen years of her life, before 1900, she was an ordinary deep-sea cargo-carrier. Now she was commanded by a youth aged 24, a cheerful young man from the captains' town of Ilhavo, a product of the new seafaring Portugal, with high qualifications from the lyceum and the nautical school, and a cruise in the training barque *Sagres.* He had made only four Banks voyages, but he came of a long line of Grand Banks schooner and barquentine masters. He had shown his ability when he deputised ably for his captain, taken seriously ill in the *Creoula* a year or two before. I had already heard something of those master mariners of Ilhavo, with their long tradition of Banks voyages. Ilhavo had been sending captains and mariners to the Grand Banks years before the island of Newfoundland was given its present name. Sailing a barquentine or a big schooner or a dory-carrying motorship on a six-month voyage to the Banks and Greenland was something any of its seafaring sons could take in his stride.

By the morning of Sunday, the twenty-sixth of March, the Lisbon fleet was assembled. Some thirty ships of the forty-five hand-liners making the 1950 campaign lay in the Tagus, all gay with bunting aloft, and their decks filled with fishermen. The fishermen had come in by the night trains from Fuzeta and from Póvoa, Oporto and Figueira da Foz, by bus from fifty fishing hamlets round the coast. The womenfolk, knowing that the day of departure was very near, had been with their men to a special midnight mass at the Church of the Jeronimos the previous evening; many had their children with them, and these, feeling the tenseness associated with all partings, were fretful, though they still loved to scamper on the decks of the ships, and to play in the riggings.

That Sunday, with the fleet at orderly anchor in the river by Belem, the special blessing service was held in the Church of the Jeronimos, where all that remains of the illustrious da Gama and the poet Camoes may lie in the stone tombs which bear their names opposite each other at the western end. Into this great church the fishermen crowded—the fishermen and the captains and the mates and the boatswains and the cooks, and the wives and the children; with them, dignitaries of church and state, ministers, admirals, officials of the Guild of the Codfishing Shipowners and its allied organisations; Commander Henrique Tenreiro, the Government Delegate on the Guild; Commander Tavares de Almeida, of the Portuguese navy, who acts as captain of the port in Davis Straits and upon the Banks for the Portuguese vessels; the captain of the *Gil Eanes*, the hospital and assistance ship, with his officers, and doctors; new boys from the *D. Fernando y Gloria* (whose stately old teak hull lay at the head of the line of ships out in the river, like a big black swan leading a brood of sleek geese) and the Fishermen's School; contingents in uniform from the navy, army, and merchant navy; the captain, officers, and crew of the French frigate *L'Aventure*, which was on a courtesy call at Lisbon before sailing for the Banks to look after French trawlers there: and, with all these, socialites, townspeople, strangers. Lisbon intended that her fishermen should know that the good wishes of the city went with them, as well as the blessing of the Church.

But it was, first and foremost, the fishermen's day. A hundred of them, dressed in their bright check shirts, striped trousers, and sea boots, stood in two strong lines across the roadway as the dignitaries arrived. They looked upon the pomp and braid, the brass and the occasional feminine loveliness, but they already saw waters far away—restless, wind-torn waters, harassed with pack ice and 'bergs, where all too soon they would be, earning a hard living in their small dories. They could see the masts and the bunting of their ships in the river from the church door, and an east wind was blowing, to take them to sea. The very church itself reminded them of their ships, for it is built like the inverted hull of a big vessel and much of its embellishment had its *motif* in the profession of seafaring.

Inside, the crowded church was flower-laden and the high altar gracious with lighted candles and lovely blooms. The Blessing was given by His Grace the Archbishop of Mitylene, Don Manuel Trindade Salgueiro, himself the son of a drowned Ilhavo fisherman, whose family was well known to many a captain and mariner standing there. After the mass he spoke to the fishermen, upon whom he asked God's blessing, in a few simple words. He did not need to remind them that he might well have been one of them. They were, he said, men who knew the true meaning of duty, to God, to the State, to family and to ship. They would voyage under God, and to God commit themselves and their families: for them all, he wished a safe and prosperous voyage, and a quick homecoming. He spoke quietly, but his words were clearly heard through the huge church, which was thronged by thousands. When his oration and the service were finished, he went in solemn procession to the great door, the stately group led by a simple fisherman carrying a cross of flowers. From the roadway beyond the door, the Archbishop gave his blessing to the ships. It had rained a little earlier in the morning, but now the sun shone upon the freshness of spring, and the old grey stones of the Jeronimos were touched to warmth, while the east wind sighed softly in all the riggings.

So the service was over and the throngs passed on. Two fishermen took a gift of flowers to the Prime Minister, Dr. Salazar, and offered him the good wishes of their brothers in all the ships. That afternoon, the scene by the Belem waterfront was animated, with a host of dories passing between the beflagged white ships and the Belem steps, and a race of yachtsmen's dinghies with coloured sails threading through the fleet, against the swift Tagus tide. Women and children were prominent in the dories, some of them bringing a little hand luggage ashore. On the stone steps by the ferry landing where the dories came and went, fishermen passed in a constant stream. Many were carrying a large demijohn in one hand, and a conch shell in the other. The demijohn I could understand, for this often contained a family gift of special wine, for feast days and special occasions. Just as often it contained boiled oil to dress oilskins. But the conch shells—what were they for? I saw a crowd from the Algarve reporting to join their ship, going out in dories which carried the sign of the cross. These men had with them their simple bedding and sea stocks of clothing in grips made of soft basket stuff and shaped like an old-fashioned carpenter's toolbag. Each carried a sail, a rolled-up oiled sail, across his shoulders. Many of these sails were emblazoned with painted banners, crosses, and devices. Why were they joining ship with their own sails? Other strange things they had with them—thin poles with rowlocks spliced into one end; curious rubber rings, about twice the size of a man's wrist and shaped like half a very small rubber tyre; tightly-lidded wooden baskets, obviously water-tight, and painted in the brightest colours and patterns, some with vertical stripes of red and white, some all golden, some purple and yellow, others a flaring crimson. Many of the men had what seemed like bracelets of gold chain on both wrists, bound round

THE FLEET ASSEMBLES

three times, though they did not look affluent. What could be the meaning of all these things?

I watched this interesting scene for hours, and wondered why these hundreds of stalwart men chose to change their sunny Portugal for the wet, cold mists of the horrible Banks and the ice-littered treachery of West Greenland. Many of them had not spent a summer in their homes since they were children, for the same men went to the campaign year after year, generally in the same ships. How could they bring themselves to leave the pleasant fishing places of the Algarve, the artists' paradise of Nazaré, the well organised fishing community of Póvoa de Varzim? And two score and more other sunlit spots?

Well, there they were, and there were the ships. And there was my berth in the *Argus:* I was going, I hoped, to find the answers to these queries. I had spent some weeks already among the fishing communities in Portugal, almost from the Spanish border in the north to Faro and Fuzeta in the south. I had been to the captains' town of Ilhavo, had listened to the captains assembled on the pier by the Monica yard at Gafanha, speaking gloomily of reports of frigid cold and too few fish, from the Portuguese trawlers already on the Banks.

"This year we are going too early," some of them said. But they went.

The fleet was led to sea from Lisbon by the Viana do Castelo ships, the *Santa Maria Madalena, São Ruy, Santa Maria Manuela,* and *Rio Lima.* They went with much wailing of the great fog sirens which are fitted in the mizzen tops or the jigger crosstrees of all the ships. There was much flying of "Bon Voyage" signals in hoists from the assembled ships, and the crowds of men on the dory-filled decks looked towards the land, and waved farewells until they could no longer be seen. This vanguard went to sea on the evening of March the twenty-seventh, and the *Argus* was to follow in a day or two.

Other ships were sailing daily from Oporto, Aveiro, and Figueira da Foz. Soon the thirty-two sailing-ships and the thirteen small motor-ships of the 1950 fleet would all be on their way, with their three thousand fishermen and all their dories, to wrest their livings from the depths of the waters for half a year and more, as fishermen had done before them for centuries, with their faith and courage to sustain them, and their ability and fortitude to see them through.

She Had a Gloucester Bow

In Dry Dock Her Good Lines Showed

Loading Salt

The Fleet Assembles in the Tagus

The Nests of Dories

Beach Fishing Type, Caparica

The Last Barquentine

An Old-timer on the Way to the Banks

CHAPTER TWO

THE BACKGROUND

> I saw these things, which the rude Mariner
> (Who hath no Mistresse but Experience)
> Doth for unquestionable Truths aver,
> Guided belike by his externall sence:
> But Academicks (who can never err,
> Who by pure Wit and Learning's quintessence,
> Into all nature's secrets dive and pry)
> Count either Lyes, or coznings of the Eye.

"THE women and children ashore now," said Captain Adolfo, very quietly as he came over the side. He was a slight, dark man with a fine face which looked rather melancholy just then, at the setting out on so long a voyage. He was dressed in a shore-going grey suit, with a soft felt hat on his thick black hair. He looked at least ten years younger than his fifty-two. Ashore, there would have been nothing to distinguish him among the Lisboetas as a Greenlands schooner captain (and his schooner the best in the fleet). But as soon as he put his feet over the rail it was obvious that the master was aboard. The ship, lying ready with all sails bent and all gear clear for running, came to life, not with any babel of shouts and noise though she lay then at foul moorings in a crowded anchorage, and it was approaching night. Quietly and efficiently she was no longer an inanimate thing of wood and steel and four high masts which men had built and rigged, with graceful decks well filled with men, and little red dories. She was a deep-sea sailing-ship, bound upon a long and difficult voyage, eager to go.

The women went sadly over the rail and were rowed ashore in the last dory. There were five or six, and they had little children with them. They were the sailors' wives and families: the main body of fishermen had joined only a day or two before, and their wives were at home. The dory went very slowly across the strip of blue water. The wives and children looked back, not waving, silently, as if appalled at the enormity of the approaching parting; and they did not wave until they were come to the stone steps of the ferry landing, where they lined the balustrade. The dory was hoisted inboard. The pilot was aboard, and anxious to be on his way. Last minute farewells were made by brother captains, the pleasant young man from the last barquentine and a tall, robust, handsome fellow who looked no more than thirty-five but said he was about to make his thirtieth Banks voyage. Silvio, they called him, Captain Silvio, and I thought that was his only name: Captain Silvio of the motor-ship *Elisabeth*.

He was a great crony of our Captain Adolfo, and we were to see much both of him and of the youthful master of the barquentine. These were stalwart fellows, and they seemed cheerful as they wished our Captain a good voyage and good fishing on the Banks. The schooner *Creoula* was also sailing with us and the *Elisabeth* would follow in a day or two. Meanwhile *Elisabeth,* moored a cable or two astern of us among four four-masted schooners, six three-masters, and several other dory-carrying motor-ships, was entertaining the assembled fleet with gramophone music played loudly and unceasingly over her large loud-hailer. Several schooners near the *Elisabeth* were still loading salt from the pretty Lisbon river-barges. All were flying "Bon Voyage" signals for us and the little group of ships putting out in company.

Further upstream, the gaunt chimneys of Lisbon's waterside power station towered over the masts and yards of the small white barquentine, and I wondered that such a ship could still find useful work to do in these days, for the power station seemed to mock her, belching its smoke upon her fragile yards. Two 20,000-ton liners in the Rio trade passing outward bound with bull roars upon their sirens, pushing fat rollers of the shining Tagus before their bows, heedlessly, like elephants, seemed to add to the threat: but the little barquentine danced gently in their wakes as she had danced for half a century before they came, and might well dance upon the friendly sea for years after they had gone again. For she belonged to the sea; and in a sense, they did not. She was in the true line of succession from all the great ships which had started from here, from this anchorage by the Tower of Belem, and all the ships that had sailed upon the seas since sailing began. It was the liners which were menaced by our days, for the barquentine and her kind could survive and work even if the power stations and all the complex world they serve were to crumble in the dust.

The sailors picked up the moorings quietly, without fuss, and we headed out to sea. The fog sirens of the fleet wailed for us—a strange way of showing good fellowship, but it was a custom of the Bankers. A short-voyage trawler, inward bound, wailed on his steam whistle; and when we had crossed the bar, the Lisbon pilot steamer wailed for us, too, as he took off our pilot.

As we came down the river, the gaunt man standing at the wheel was sometimes in tears. He was a fine figure of a man, lithe, and splendidly built. A beret was jammed upon his head. His deep-sunken brown eyes, though softened now with tears, were fierce, and determined. He looked like an athlete and he had the stance of the true sailor. Steadily he followed the pilot's orders, with a few spokes this way and that.

"That man is the First Fisher of Portugal," said the young mate, buttoning his brown battle-dress round him against the evening breeze.

"'E makes feesh," said the second, who looked even more youthful. What a First Fisher was, I had then no idea, but I was given to understand that our helmsman was an expert so skilful at the art of catching cod that none excelled him in that land, and indeed he did so well that the others suspected him of having private hatching-grounds somewhere off Greenland. I looked at the gaunt man with fresh interest. Certainly there was something striking about him, whatever his secrets with the cod; and I could see he was not thinking about fish then.

It was the night of full moon, Saturday the first of April, 1950. We gave her all sails in the bay off Cascais, and ahead of us the *Argus*' sister *Creoula*, under all plain sail, was a picture of white loveliness, gentle and graceful beneath the moon. The lights of Portugal fell rapidly astern, and we wondered when we should see them again. We passed the *Creoula*, treading the white

THE BACKGROUND

water down at her curved bow, and Captain Adolfo's brother Captain Almeida shouted an evening greeting. The wind blew quietly, and the ships, deeply laden with salt and stores, heeled under a press of sail until the cold water of the North Atlantic began to break aboard. The moon shone brightly on the long maindeck, gleaming with the sea: a group of the mariners on watch gathered by the binnacle aft, and the green and red sidelights sent shimmering reflections upon the sea. The First Fisher of Portugal had been relieved: the man at the wheel now looked more than sixty. As he came to relieve the previous helmsman, he touched his forelock and, in a clear voice, recited some formula in which all I could understand was the name Jesus Christ. This done, he mounted the grating, and firmly grasped the spokes.

Down for'ard in the "rancho" where the sailors and the fishermen lived—these were the same men—the three tiers of wooden bunks groaned with the weight of their occupants, stretched in fitful slumber. I had a good cabin aft, the North Atlantic before me, and all the time in the world.

Historically, it was fitting that the Portuguese should be the last to send a fleet of sailing-ships across the North Atlantic. Long-voyage navigation in that wild ocean was pioneered by the Portuguese and the Norwegians. There is plenty of reason for believing that Columbus was preceded over the Western Ocean by a number of truer pioneers, and that these pioneers included a host of simple fishermen sailing from Portugal and the Basque Biscayan ports, to the Grand Banks off Newfoundland for cargoes of cod. None of these ships ran the down-hill way before the trade winds, because they knew where the cod were, and they were not to be found in warm waters. It was Columbus's good fortune that he stumbled upon the outlying islands of a great continent when he thought he had reached Asia. For the rest of his life, he went on thinking he had reached Asia, and for that and many other reasons, was a source of trial to those who had to work with him. There is evidence that well before the end of the fifteenth century the Portuguese (who had then been making ocean voyages in the Atlantic for upwards of half a century) knew what they wanted to know of that great ocean—that it did not offer a simple sailing route westwards to the East, but that it did possess the Newfoundland banks, which teemed with cod.

Like the Portuguese, the Norsemen were intrepid sailors and expert fishermen. Their country was small, mountainous, and poor. It had evolved a race of men hardy, of necessity adventurous and, equally of necessity, familiar with the ways of the sea. They were fish-eaters: their Arctic coastal waters abounded in cod. They understood the catching and drying of cod very well. In Iceland they found a further excellent supply of cod, and in the course of time they built up a trade in dried cod, called stockfish, to Ireland, the West of England, and elsewhere in Europe. Cod had the great merits of abundance and of lending itself readily to preservation by the primitive methods then available (which still serve). In an age when the preservation of most staples was impossible, and, generally speaking, meat was a luxury for the rich in summer and unobtainable in winter, stockfish were gold, and the Norsemen had something like a corner in them.

THE BACKGROUND

From Iceland to Greenland was a logical step for hardy seafarers with good ships, and so was the next, from West Greenland to Labrador, Newfoundland, and Nova Scotia. The Norsemen knew three separate areas of the American continent as Helluland, Markland, and Vinland. There is understandable argument about the actual identity of these three places, but it seems likely that they were Labrador, Newfoundland, and Cape Breton Island. But Norse expansion in this direction was discontinued, very probably for two reasons. One was the warlike opposition of the aboriginal inhabitants, and the other the fact that these areas were too remote from the motherland to lend themselves to successful colonisation. The Vikings already had abundant sources of timber and cod. They turned their attention to richer countries nearer home which were already well settled and could be raided more or less at will.

The Portuguese pioneering voyages in the North Atlantic were in a different category. In Prince Henry—called "the Navigator" though he never actually sailed a ship—Portugal had a directing genius of exceptional vision and determination. He set out to make his people into a great maritime nation, and it was at his instigation that, in the second decade of the fifteenth century, Portugal entered upon a deliberate policy of ocean discovery, as opposed to one of land journeys. One of the objects was to secure a sea route to the markets of India and the Far East generally, and so to attack the wealth and power of the "Moors" at their source. These markets were well known and had been visited by overland routes. The theory that they might also be reached by sailing westwards across the Atlantic Ocean had been put forward by the Greeks, among others, and attempts may have been made. Coins of ancient Carthage have been found in the Azores. Aristotle said the world was small, and Gibraltar not far from India, an airy statement and all very well for a philosopher in Athens. But it had the germ of an idea, and Henry the Navigator, early in his career, set out to investigate the North Atlantic. One of his first steps was the discovery, or rediscovery, of the Azores, which were on the maps by 1439 and were probably known in Portugal at least a decade earlier. These were a convenient base for further Atlantic discoveries, and they were easily reached from Portugal by sailing-ship, for they were only about 800 miles away. From the Azores to Cape Farewell is less than 1500 miles, and to Cape Race, little over 1000 miles.

In order to make voyages over these distances, it was necessary to possess two things: first, suitable ships; second, a nautical science. Prince Henry developed both. The Portuguese caravel of the fifteenth century was a remarkably seaworthy and weatherly vessel. The Portuguese was a mariner accustomed from his childhood to contend with the North Atlantic across open beaches, fishing in small barks and flat-bottomed open boats which had to be ruggedly constructed and of great seaworthiness. He knew his enemy the sea, and so did his shipwrights, who served him well. As for the nautical science, the great contributions of the Portuguese in this field are generally acknowledged.

King Eric of Denmark married a cousin of Henry the Navigator, and there was traffic between the courts. One Abelhart, a Danish pilot, came to Sagres by Henry's invitation and it is reasonable to assume that he discussed with the Infante the traditions of the Viking

THE BACKGROUND

voyages. The Danes were interested in Iceland and Greenland and sent at least one expedition to South Greenland during the fifteenth century. The significance of the old Viking voyages to Prince Henry was their discovery of the open water of Davis Straits. Where did that lead? Was there a way through towards the great East, as some said? But ocean voyages cost a great deal of money. Portugal was small and poor. The revolutionary idea of opening a sea route to India had to be hidden from possible rivals and from existing traders with the East who were profiting from the fact that no such route then existed. The Portuguese had to move warily, and here lies the principal reason for much of the obscurity which still clouds the story of the early pioneering voyages in the North Atlantic.

From 1440 onwards, Portuguese ships were roaming the Atlantic Ocean; but apart from odd scraps of information, we know very little about them except those which voyaged to Africa. Voyages of discovery were not publicised at the time, and their results were known only to those entitled to know about them. No discoverable records exist of many of them. But here and there, odd references throw light upon what was being achieved. Two such scraps of evidence speak of successful transatlantic voyages, the first made by Diogo de Teive, about the year 1450, and the second by João Vaz Corte Real and Alvaro Martins Homem, in 1472. Both set out from the Azores: both crossed the North Atlantic. It is the Portuguese belief that the first opened the fisheries on the Grand Banks, and that the second discovered the American Island which was later given the name of Newfoundland.

With Diogo de Teive were his son João, and Pero or

THE QUEST OF THE SCHOONER ARGUS

Pedro de la Frontera, a pilot, said to be a Galician. There is much confusion in the scanty records of this voyage. It is not even clear whether they sailed northeast (which would be senseless, for the route towards Ireland was well known) or northwest. In a notebook in which he carefully kept all the records of Atlantic voyages he could find, Columbus noted the de Teive voyage. That notebook has, unfortunately, been lost and our knowledge of its contents comes from Las Casas. Las Casas says that the de Teive caravel departed from Fayal, "and they sailed to the northeast so far that they had Cape Clear towards the East, where they found winds to blow very brisk and the winds westerly and the sea to be very smooth, which they believed should be because of land which should be there, which sheltered them from the westward: the which they did not follow up to explore, because it was already August and they feared Winter."

I think there is no doubt that the course northeast should read north*west*. How else could Cape Clear be to the *east* of them? After six months in the *Argus*, I was still unable to tell when the helmsman was giving the course as northeast or northwest, so confusingly alike are the Portuguese expressions for these opposite directions. If Diogo de Teive and his party found themselves in smooth water, with the wind blowing west, and the Irish Cape Clear was to the east of them, then they were under Nova Scotia or Newfoundland. It is quite feasible that they could have been there and have seen nothing of the land. To this day, many ships find themselves in much the same predicament.

Soon after the de Teive voyage, Portuguese fishermen were sailing quietly from the ports of Portugal

THE BACKGROUND

and of the Azores, bound for those Banks where the cod abounded. Shortly after them came the Basques, and the Bristol mariners of the west of England were not far behind. None of these activities was given any publicity. According to the Spanish ambassador in England at the time, Bristol was equipping caravels annually between 1491 and 1497 "to go in search of Brasil and of Antillia." The "Brasil" they sought was a nice cargo of fat cod from the Grand Banks, and they knew where to find it. Bristol had long been a center for the stockfish trade from Iceland, and Portuguese merchants frequented the port, taking part in this trade. By the time the older Cabot—another Genoese who, like Columbus, had been in Lisbon and in Seville—crossed the Western Ocean, knowledge of this new source of golden cod must have become fairly general among fishermen in the stockfish trade.

After the Cabots, it is significant that the King of England turned directly to Portuguese to continue such voyages, when he sent João and Francisco Fernandes and João Gonçalves of the Azores, apparently to form a settlement somewhere between Greenland and Florida. Almost nothing is known of the accomplishments of these navigators, but they must have been to the King's satisfaction, for they were granted Ten Pounds from the King's purse. These grants were made in 1502.

1502 was two years after the public formation of the first Guild of the Codfishing Shipowners, which was formed by the Portuguese ports of Aveiro, Viana do Castelo, and Angra in the Azores. By 1506, the King of Portugal was ordering the fishermen of Portugal "on their return from the Land of the Bacalhau to pay a tenth of their profits" to his Customs, and this might refer not to their profits on cod (for cod was already an old item on the country's import lists) but to the profits they had made by free trade on the Banks and in the harbours of Newfoundland. It was not long before the astute business men from the west of England, profiting by the discovery of a haven where no monarch ruled and there were no restraints of trade, were using the Grand Banks voyage not only to catch cod. A free trade sprang up in all sorts of things, which were exchanged among the ships, and merchants went out with the ships to build up and prosper on this commerce. Many fortunes in the west of England were founded in this way, and news of this trade, surreptitious as it was, would have reached the King of Portugal.

The connection of the Azorean port of Angra with the trade can be traced directly to João Vaz Corte Real, who was given the captaincy of Angra in Terceira as part of the royal reward for his successful voyage. The success of this voyage is sometimes challenged on the grounds that insufficient evidence has been discovered to support it. Considering the nature of the times and the purpose of such voyages, it is remarkable that any evidence should have survived at all. It very probably would not have done so, had it not been for the efforts of Gaspar Fructuosa, the industrious chronicler of the early history of the Azores. In 1472, Fructuosa says, João Vaz Corte Real and his companion Alvaro Martins Homem landed at Terceira, and they had "both arrived from Terras de Bacalhau,* which they had discovered by order of the King."

* Bacalhau—pronounced buckle-yow—is the Portuguese word for cod.

THE BACKGROUND

The point is that these fifteenth century pioneers, the Corte Reals and the de Teives, Fernandes Lavrador and all the rest of them, are not to be treated as straws in an academic breeze, or as figments of the Portuguese imagination. In the great maritime history of Portugal, there is no need to imagine voyages. There are not wanting other fascinating pieces of evidence in support of these Portuguese voyages—old maps, globes and odd references. The very name of Canada could well come from the native repetition of the Portuguese "Aqui nada!" shouted by some disappointed enthusiast looking for gold. Aqui nada! (Here is nothing) easily becomes Keenada. Though scarcely flattering to the Canada of today, the name may have been derived in just this way.

A more serious piece of evidence is to be found in the remarkably able and accurate sailing directions given to Cabral for the voyage during which he made the official discovery of Brazil, a country which may have been known to the Portuguese even before the middle of the fifteenth century (the coast is shown on a map drawn by Andrea Bianco in 1448). These directions could scarcely be bettered even now, though for centuries we have been accumulating data on the subject of trade wind voyages and ocean sailing. On the face of it, it seems completely impossible for such able directions to have been prepared at a time when the only knowledge of the North Atlantic was the little gained by the Guinea voyages, the trade to the Azores and Madeira, and the great passages of Bartolomeu Dias and Vasco da Gama. That kind of knowledge could have been acquired only by half a century of more or less unrestricted Atlantic voyaging. From 1440 onwards, the deep-sea sailors of Portugal were taking brave caravels manned by good mariners and fishermen down the coast of Africa, to the Azores and the Newfoundland Banks, into Davis Straits and towards the coast of Florida, from Finisterre to the Gulf of Guinea, through the Doldrums and the Sargasso Sea. They could not do these things by coasting. They had to understand the Atlantic wind system.

Here and there, a tiny corner of the veil has been lifted sometimes only to increase our puzzlement—the discovery of the name of Miguel Corte Real, a son of João Vaz Corte Real, carved with the arms of Portugal in the Dighton Rock by Narragansett Bay, not far from Boston, though Miguel Corte Real has been missing from a Davis Straits voyage for 450 years; the unexplained statement of the pilot Duarte Pacheco Pereira, one of the Portuguese delegates at the negotiations for the Treaty of Tordesillas, that he knew of the land mass on the western side of the North Atlantic which extended from 70 degrees north latitude at least to 28½ degrees of south latitude, and probably much more; the many inscriptions on old maps showing Newfoundland as the "Land of the Bacalhau" and the "Territory of the King of Portugal," before the dubious claims of the elder Cabot to its discovery were ever put forward.

Columbus did not go to Davis Straits and he avoided all Labrador, Greenland, and Newfoundland like the plague. Since he was also so loath to cross the equator that he never once sailed south of the Line, his westwards voyage towards India brought him by chance upon the discovery of the exploitable Americas. Leaving the latitude of Southern Portugal, going south a little for the true trade wind and then running west be-

THE BACKGROUND

fore it, he could scarcely go wrong. His achievement was remarkable and its results were incalculable: it is no detraction from it to offer some honor to the truer pioneers, or to recall the debt of navigators to the courageous Portuguese mariners who first dared to cross the wild Western Ocean regularly, in small vessels, seeking and finding cargoes of the homely cod.

As I stood on the heeling decks of the schooner *Argus* I was glad that I had learned something of the five centuries' background behind this 1950 Grand Banks campaign; and I looked about me with greater interest. Captain Adolfo and his mariners were in the true line of descent from the pioneers. Five hundred years! The history of Portuguese Banks voyaging in that long time is a chequered one. There have been ups, and there have been downs—ups when the industry was thriving, downs when far too great a proportion of the country's needs was imported in other vessels, and too many Portuguese vessels were withdrawn from the trade. Fishermen and fish merchants were always fiercely individualist. One or other of them brought the industry to the verge of ruin, more than once. It was generally the fault of the merchants. It was far too easy for them to flood the home market with imported fish just when their own fishermen were coming home to sell the hard-won cargoes on which they were dependent for the year's living, and so to ruin them. There were other sources of dried cod, besides the Newfoundland Banks —Iceland, Norway and, for many years, the clearing-house of Bristol in the west of England. As in other countries, there was never a dearth of merchants who put their own profits ahead of the real interests of both the industry they exploited and their country.

The last of these periods of chaos in the deep-sea fisheries was in the 1920's. This came to an end with the renascence of Portugal in the early 1930's. One of the first acts of the new government was to form the Gremio, the guild of the codfishing shipowners, to control and regulate the industry. The fish merchants and the fishermen still rated individual freedom—or their conception of it—above all other considerations. All agreed that reorganisation was essential, if the Banks fleet was to survive. But the only way in which they could be made to accept the necessary reorganisation was by compulsion. The owners *had* to join the Gremio, the merchants *had* to accept organised marketing, the fishermen *had* to submit to direction and control.

The gremio of the codfishing industry, like similar organisations in other sections of Portuguese life, had two main aims—to ensure that the industry provided food for the Portuguese people at a reasonable cost, and a living to those engaged in it. It assisted in the development of a modern fishing fleet, and it set about immensely improving the living standards of the fishermen both afloat and ashore. It was charged with the direction of fishing, drying, and the sale of cod and cod by-products. It aided in the establishment of mutual insurance groups; it regulated labour contracts and put an end to unsatisfactory practices in this field; it provided adequate medical aid for the fishermen afloat and for their families ashore. It set about developing new skills in fishermen, and the retention of the best of the old. One means of achieving this was by the establishment of fishing schools at centers round the coast. Either directly or through associated organisations, it undertook a great rehousing programme for the fishermen ashore, and its lovely settlements of fishermen's houses are now a feature of every fishing port and center of importance round the coasts of Portugal, at Madeira, and in the Azores. Many of these settlements are already of more than a hundred houses, some of almost two hundred, and there are plans for thousands more such homes. Fishermen's Institutes, schools, dispensaries, medical centers, refectories, recreation centers, even maternity hospitals for the fishermen s wives and crêches for their children, are now features of life in the lovely coastal towns. The directing genius behind the Gremio, realising that much could be achieved for the fishermen through social welfare, an aspect hitherto almost entirely overlooked, has in a remarkably short period done wonders. But that is only one of the many achievements of the Gremio.

The beautiful schooner *Argus*, old in a great tradition and new in perhaps a greater, was also an expression of the Gremio's aims and activities. In her were combined the best of the old and of the new: she was a sailing-ship, but also a full motor-ship. She fished by dories and hand-lines with hooks and bait, but her dorymen lived in steam-heated quarters, drew bait from a refrigerator, and were guarded by the boon of radio telephony. She was to be six months at sea, but her crew of family men were kept in touch with their people by weekly radio broadcasts and their mail was brought to them, even on the Greenland grounds, by the Gremio's hospital-ship. The *Argus* was a sailing-ship in true line of descent from the caravels, the luggers and the schooners which had preceded her in that hard trade for perhaps five centuries, more lovely even than they and stronger and more seaworthy; and her people would not be plagued by the gnawing worry of ruined markets for their cargo, when they returned.

Behind this progress were men, not many men. The Prime Minister, Dr. Salazar, his responsible ministers, who included Dr. Pedro Theotonio Pereira, and the government delegate on the Gremio and for all fishing activities. This last was the dynamic Commander Henrique dos Santos Tenreiro, untiring and indefatigable, a genius for organisation. Now the pioneers had moved ashore; but Commander Tenreiro and his kind, like the ships they served, maintained the great traditions of the past. So did the owner, Senhor Vasco Bensaude. Without the genius of such men, there would be no *Argus* to spread her white sails gracefully upon the Atlantic sea, no fleet of sailing-ships to fly westwards before the east winds of spring.

Modern Portugal is indeed fortunate, both in her ships and in her men.

CHAPTER THREE

ACROSS THE NORTH ATLANTIC

> Set sayl (he cride), set sayl to the large wind:
> Heav'n is our guide, and God our course directs.

ON THE fourth day after sailing from Lisbon, the *Argus* arrived at Ponta Delgada on the island of St. Michael's in the Azores. Wind and weather had been favourable, and the passage without incident. The *Creoula* left us off the breakwater of Ponta Delgada, and continued towards St. John's. Her crew was complete. Of the whole fleet, only the *Argus* and one other schooner, a three-master named *Oliveirense,* were to engage fishermen in the Azores. It is unlikely that the *Argus* would have touched at Ponta Delgada, had not her principal owner, Senhor Vasco Bensaude, taken a real interest in the welfare of its people. Even from the buoys behind the breakwater, where the schooner was moored, it was obvious that St. Michael's lovely green fields had been parcelled out among the inhabitants to the utmost possible degree. The island was over-populated, and there was some poverty. No smoke was rising from the factory chimneys round the pretty port. Azoreans have a long tradition of successful deep-sea fishing and whaling. In bygone years, these islands and the Cape Verdes provided many an expert crew for a New Bedford whaler, and the movement between the Azores and Massachusetts was on such a scale that, today, scarcely a family in the islands does not boast some relatives in America.

We could have recruited two hundred dorymen there with ease, but we needed only twenty-six, and two deckboys. I first saw some of our fishermen at a little village near the eastern end of Ponta Delgada. They were standing in a group, barefoot and carrying staves, singing outside the door of a church, and while they sang, the church door opened and they went in, still singing. They left their staves in a neat stack on the porch. Their strong faces looked peaceful and serene, and their chanting was melodious and moving, in that village square. I was told they were giving thanks to God, to whom they were commending themselves before setting out on the voyage. They looked poor and hardworked and, by our easier standards, not men with a great deal to be thankful for; and they were about to leave for a period of many months the families to which they were greatly attached.

The next day they came aboard, now gay in their brightly checked shirts, and boatloads of friends and relations came to wish them good fishing. They passed up their bedding and their small bundles in silence. It was blowing fresh, then, and had begun to rain. We were to sail as soon as the embarkation was complete, for we had no other business in the port. But the wind began to roar down the green valleys and over the mountains, savagely, and the sea rose inside the harbour. It was soon as much as the ship could do to keep her place, going full ahead on the motor with the two anchors down. How it blew! The wind whipped up whirlpools of powdered spray which it flung impetuously across our decks, and there was such strength in it that it was unsafe to attempt manoeuvers in the confined space of the harbour.

In the morning the sun shone and the wind had gone to the north-east, fresh still but not more than that: we could go. We went out with a wail of the great siren in the jigger-top, and two boats full of men and boys rowed hard to keep up with us, waving farewells as long as they possibly could. We turned westwards off the breakwater's end and passed close by the land, making plain sail. The fishermen's families lined the hills, waving frantically. Our Azoreans waved back, and Captain Adolfo blew a salute on the mighty siren. Dogs ran, barking. A small boy on a bicycle followed the ship on a rough track ashore, as long as the track lasted. Most of the dorymen looked moved—there was one, they said, who had been married only the previous day—but they soon forgot the passing grief of parting in their pleasure to have the opportunity to earn a few thousand escudos, for escudos were scarce among the islands.

Once clear of the land the wind was strong and the sea turbulent, washing over the low waist and smashing at the nests of dories as if it wanted to hurl them away. The ship rolled and leapt along under the four lowers and two headsails, making a good ten knots. Captain Adolfo was eager to catch the *Creoula* again,

THE QUEST OF THE SCHOONER ARGUS

if he could: in the afternoon and early evening he was calling her on our radio telephone. The captains used the telephone to provide a picture of wind and weather across the North Atlantic. Strung out now between the Banks and Tagus Bar were a score or so of the big fishing schooners, the westernmost already in touch with the trawlers on the Grand Banks (which also used the radio telephone) and the easternmost still within sight of the coast of Portugal. Each passed its weather news to the ships in immediate touch, and in this way all knew what was brewing. We could make the best use of the wind, and bowled westwards merrily. Even the schooners without auxiliary power had radio telephones and small motors to provide power to operate them.

As soon as the main-deck was reasonably dry, the preparations for dory-fishing began. The twenty-six Azorean dorymen were mustered aft, a silent group of dark and serious men, many of them over fifty years of age, perhaps more than half of them. They were given their free issues of thigh sea-boots, a sou'wester apiece, six pairs of woollen gloves, half-a-kilogram of smoking tobacco, and several litres of oil to prepare their oilskins and to oil their dory sails. Each doryman was entitled to a free issue of eight litres of oil for the campaign, and to make so generous a distribution the *Argus* had to carry large quantities.

The *Argus* was carrying large quantities of many kinds of things needed for fishing. First were the dories, each of which was distinguished by a large numeral, painted in white on both bows. For these, as part of their necessary equipment, were 200 boat anchors, 300 oars, 100 masts, 100 wooden bailers, 100 wooden pails for the bait, 60 boat compasses (which were distributed last), 100 baskets to hold the long-lines, 300 sharp knives and 100 honing stones (being one for each doryman, and the necessary spares), 300 yards of duck for sewing dory sails, 1000 lengths of stout cod-line in fifty-fathom pieces and another 500 coils of lighter line to make the "snoods" (or subsidiary lines) to which the hooks were attached, and 100,000 hooks. There were also stocks of large iron spikes for making grapnels for the long-lines, and forged stocks into which the spikes were to be fitted; half a ton of lead for sinkers; hundreds of fish-jigs, and a thousand swivels for the hand-lines; 40 gaffs which were light pitchforks, very sharp and kept surgically clean, for pitching heavy cod aboard from the dories, and handling them on deck; large stocks of timber for making thwarts and portable bulkheads, to prepare adequate fish stowages in the dories; small sheaves set upon steel holders, the better to control the lines; scores of pairs of rubber gloves for the fish-salters to use, when their hands were chafed and cut by the tough fishing-lines; rowlocks, buckets, barrels, drums, cordage, spars, paint; large vats for fish-cleaning, and steel and copper piping, and brass taps, to provide a running sea-water system to clean fish on the grounds.

So also with the foodstuffs—the supplies were enormous. There were barrels of wine and brandy (wine was issued to the men on Thursdays and Sundays, and in the saloon every day) and of olives and olive-oil, as well as sugar, flour, and beans and peas of all kinds. There were casks of salted pigs' trotters, and Argentine boneless beef; cases of canned goods, and even of vitamin pills; drums of lard, sacks and cases of potatoes by the ton; large supplies of macaroni and similar products. Both refrigerated rooms, which were immediately abaft the galley, were stowed to capacity with fresh food. A compartment aft was well filled with medical stores, and spares for the seven motors—one main diesel engine, and six auxiliaries for the lighting, refrigeration, windlass, and pumps—and electrical installations, which ranged from echo-sounding and radio-telephony to a complete loud-speaker system, had another compartment.

The *Argus*, although essentially a seagoing dormitory for the fifty-three dorymen, a stowage for their dories and their fish, and a depot from which to conduct fishing operations, was obviously a complex and unusual vessel, the more so because of the apparent simplicity with which she was controlled. There was a free democracy fore-and-aft. All hands had the run of the ship, though some lived for'ard, and some aft. The watch system ran so smoothly at sea that the changing of the watches by day was scarcely noticeable. The decks were crowded with men while there was work to be done, which was as long as the daylight lasted. All the dorymen were on deck, and at work, all day, and by night they served in three watches of four hours each, one watch the mate's, one the second mate's, and the third the boatswain's, who kept watch for the captain.

The first task of the dorymen was the preparation of their dories, and the first step in that was to decide which dory they were going to have. All were alike; one was as good as any other, as far as I could see. They were all little red punts with flared-out sides to give them good stowage room for gear and cod. The traditional method of deciding who should have which was by lots, and a small canvas bag hung ready in the charthouse, containing little blocks of wood each of which bore the number of a dory, but there was none for Dory 13. Captain Adolfo brought the bag on deck and the dorymen filed past, each drawing a block, while the mate entered the numbers drawn on a list. Then the nests of dories were torn apart; the twenty-six which the Azoreans had drawn were shifted to the port side, and the Mainlanders' were stowed on the starboard side. This was to be the permanent arrangement—Azoreans to port, under the mate, and the Mainlanders to starboard, under the second mate. This marked no schism between the Islanders and the mainland Portuguese, who were the best of friends. It was a convenient manner of working the ship. Down for'ard in the big forecastle—always called the rancho—where they lived, there was no such division. The sixty-three men and boys who lived there took bunks from the seventy-two down there as they wished, though brothers and fathers and sons usually preferred bunks close to one another. There were two fathers and sons, and five pairs of brothers: the cousins, brothers-in-law, sons-in-law, and other relatives, I never counted.

The main deck was soon strewn with dories, each with at least one man busy in it, shaping thwarts, rigging leads for his lines, adzing plugs to fit the draining-holes, looking to the mast-steps, and fitting wooden bulkheads to keep the fish from his feet. Every man was a carpenter but there was neither a shipwright nor a carpenter's shop aboard. There were plenty of tools. Each man held his piece of timber in one hand and adzed, chiselled, or sawed away at it with the other. Some blood was spilled, inevitably, but this did not stop the work. Often the timber was a piece of a tree, and much of it looked like small eucalyptus saplings. This it was, for the Australian gum tree has flourished in Portugal. Saplings provided spars for the dory sails, and masts, and poles to which large hooks were secured for hauling up big fish.

The men worked cheerfully and industriously, the boatswain along with them, for he was a doryman as well. The five sailors were dorymen: everyone was a doryman, except the officers, the cooks, and the deckboys. The officers and the deckboys helped generally with the carpentering, and Captain Adolfo was the energetic and far from silent foreman of them all. He was about the decks all day, seeing that every minutest detail was arranged with perfection, and superintending the issues of dory gear. Everything issued was listed, but not signed for, because some of the older dorymen could not write. A few of these painted private marks on their dories, so that they would know them. Most of the men painted religious emblems on their dory bows, and texts and the like. It was soon obvious that their religion was part of their daily lives, and not a vague feeling largely bereft of real value, to be thought of only on Sundays. The watch at midnight was called by a doryman—Tiago Rodrigues, from the Algarve—who sang a song, standing at the head of the companion leading to the cavernous rancho. Tiago was a youngish man, the father of three fine boys, and I never discovered why he was chosen to sing these special songs, which were of a definitely religious nature. He was a well-built, dark man, with strong, well-defined features: he had been a doryman for fifteen years, and was making his twelfth voyage in the *Argus*.

ACROSS THE NORTH ATLANTIC

He would come to the head of the rancho companion sharp at eight bells and, in an excellent baritone voice, awaken the sleeping watches (regardless of the fact that only one watch was required to come on deck, and members of the other—the boatswain's—could go on sleeping). One of his songs went something like this:

> Oh praise ye the Lord,
> And adore Him always.
> Blessed be our ship
> And all her people:
> May God be our guide,
> The Virgin our Guardian.
> Rise up the watch!
> One man to the wheel
> And two to lookout.
> Rise up the watch!
> 'Tis middle of night.

That was all the call they got, but the men rolled out promptly. Fishermen, apparently, were used to being called out in a hurry. Only deckboys turned over and went to sleep again, and they were excused watches.

Going aft to relieve the wheel, the new helmsman always doffed his cap or touched his forelock as he reached the grating at the wheel.

"Praise be to God," he would say, "and to the Lord Jesus."

Then he would be given the course, repeat this—as do sailors in all ships—and take over the steering for the following hour.

Tiago the singer's dory was number 31, and I watched him as he prepared it. First, he spliced in new grommets at bow and stern. These were short lengths of rope rather like elongated deck quoits, which were rove into the bow and the small counter, and it was by means of these that the dory would be hoisted in and out, when fishing began. Then he fitted a mast-step, well forward, and after that, the thwarts, of which there were two. He contrived one to act as a bait-cutting table, with battens nailed to it to keep the cut-up bait from rolling off. On the other, he made a small rowing seat, very rough. Then he tried his mast, and put a leather thong at the bow-head to which he could secure his tiny jib (for he was one who preferred a jib-headed main and a little headsail: in some ships the dorymen used dipping lugsails) and a ram's horn on both sides of the rail aft, to which he could secure the main sheet. He spliced pieces of light line to his rowlocks, in order that they should not be lost, and fitted a sort of pulley-wheel to an iron frame forward, giving this several alternative positions, and when I asked him, he said he would always pull in his long-line over that pulley, and a 600-hook codline took some pulling in, to say nothing of a 1000-hooker. He examined all the fittings and gear which had been issued to him, to ensure that the wooden bailer, the pails, the line-basket, and the rest were in good order. He shaped a stick from a piece of gum-tree, and secured a large hook to one end: this, he said, was for hauling in cod which came to the surface lightly hooked, as cod frequently did, for the cod was a fool fish which never tried to get off the hook, unless it fell off. He made another piece of wood for jamming down big cods' throats, to hold their mouths open and facilitate the removal of the hook. He fixed a handline-holder on a swivel, so that it would run easily when he was using it. He prepared sinkers for his jig-lines, and swivels: and meantime he had also bent his sails, and tried them; and freshly oiled the main and the jib, and his oilskins, and sewed himself a large oilskin apron, which he oiled green—a bright pea-green—for no particular reason. All the dorymen loved colour. They made their oilskins and their dory-sails bright colours—reds and greens and blues—just for the sake of brightness.

While Tiago Rodrigues and the others were preparing their dories, a great deal of other work was going forward. Lines were being prepared, and in this there was no stinting of new gear. One fifty-fathom length of line cost more than ten shillings, and the best galvanised Norwegian hooks were used. Short lengths of lighter line, called snoods, were attached by special knots to the long lines, these snoods being a fathom or so apart (there was no accurate measurement: each man took his span as a rough guide). When a long-line was made up with 500 snoods or so, each of which held a sharp cod-hook, the whole thing required skilled handling if it were not to become a hopeless tangle; but the dorymen were experts at that kind of thing. With a deft flick of a strong wrist, any of them could control a hundred snoods and hooks in an instant, and they were always most particular about the coiling down of their lines.

They were, indeed, most particular about everything to do with the fishing gear. There were a few green fishermen—first voyagers—but for the life of me, I could not tell which they were. I thought at first that they were probably the youngest, but they were not. A dozen or so dorymen were in their late 'teens or early twenties, but most of these were already Banks veterans. The *Argus* crew seemed to be the same year after year, for the men were all friends and old acquaintances. Captain Adolfo knew them all by their first names. Even the deckboys were experienced seamen. Three of them, indeed, were married.

One day a small deputation of the Mainlanders and one of the Azoreans came aft and asked if they could hear the commentaries on a football match taking place that afternoon in Lisbon, between Spain and Portugal. The loudspeaker system was hooked up and a large speaker rigged forward, on the for'ard house, and within a few moments, the frantic voice of the Portuguese radio commentator was shrieking to the North Atlantic air, while the delighted dorymen galloped on with their work and followed the interminable movements of a wind-inflated leather ball, two thousand miles away.

"Good," said the mate, smiling broadly. "For work an excellent tempo, don't you think?" and indicated that one of the main reasons for installing the deck loudspeaker system was to provide just such quick-fire programmes, for the men responded subconsciously by quickening their own movements. But they worked fast enough at any time.

Captain Adolfo, whose interest in the football was about as great as mine, suffered the noise good-naturedly, and went on superintending the erection of a queer piece of minor engineering along the length of the fore-deck, on both sides. This looked like two sets of miniature railways on wooden tracks.

"For the fish," said Captain Adolfo. "Plenty fish, plenty work."

ACROSS THE NORTH ATLANTIC

There was plenty work getting ready for plenty fish too, it seemed to me.

As day followed day and still the north-east and east-north-east wind blew, we sailed steadily towards the Grand Banks, and the railway on the fore-deck took shape. Round it grew the fish-pounds, made of sections of steel and wire mesh bolted together, and with these, the running-water fish-cleaning system, which consisted in the main of five huge vats on each side, each with a large brass tap to feed sea-water into it. The water, apparently, washed the cod and then slopped out, and drained away outboard with the roll of the ship. The fish-pounds were as high as the ship's bulwarks. There were six altogether—enough to hold about 250 quintals of fish, the mate said.

"Two hundred and thirty," said Captain Adolfo.

"Two hundred and sixty," said my friend César Mauricio, the Lisbon engineer who had charge of the big diesel and all the auxiliaries, and had a man and a boy to help him. "Captain Adolfo is a first-class pessimist, when he is reckoning fish. He never allows for the heaped-up fish on top."

"Fish shrink," said Captain Adolfo.

The second mate grinned, for it was obvious that the precise measurement of the take of fish, any day and any voyage, was more or less anyone's guess. The mate explained that a Portuguese quintal of cod was sixty kilograms of *dried* fish. Since it never was dry aboard the *Argus*, but always fresh or in salt bulk, it was difficult to see how such a measurement could be used with accuracy. It was the standard. The men always spoke of "quintals." The First Fisher, for instance, was said to have taken nearly 500 quintals of cod the previous campaign. This meant that he had taken enough fresh fish to convert into 500 quintals of the dried cod called bacalhau, and as this bacalhau was less than a third the weight of fresh fish, a simple piece of arithmetic showed that the man must have caught over a hundred tons of fresh cod. A hundred tons of fish! Here was a fisherman. He wore a check shirt, check trousers, wooden clogs, and skull-tight black beret, and happened then to be oiling the wheels of a little trolley-car which was going to run on the rails. His gaunt face was thinner than ever, and his blue eyes fiercer. The men working with him were his brother, his nephew, and a protégé from Fuzeta who was a green fisherman, they said. The green fisher had the merry countenance of the born clown and seemed to be finding the idea of the wooden railway a great joke; but the others were serious. The brother was a much older man but the nephew looked about 18. He was a third-voyage doryman and an expert salter, the mate told me, one of the best.

A wooden working platform was built up over the steel hatches between the fore and mizzen masts, and all the dory nests were shifted aft to leave the foredeck clear for working fish. Gradually the scheme of the fish-handling plant began to become clear—the pounds to hold the cod, vats to wash them, working tables atop the vats to split them and gut them, trolleys to run them along to the hatchways above the hold from which they would be thrown down canvas chutes to the salters below. They certainly were preparing to handle cod in bulk.

Meanwhile a gang of the deckboys had been busy clearing the salt from a compartment in the hold, to be ready for the reception of fish. The hold was sub-

The Main Deck Was Full of Dories

Dorymen Preparing Their Long-Lines

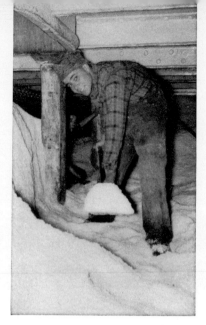

Captain Adolfo Using the Microphone. All the Ships Had Radio Telephony

A Deckboy in the Hold Shovelling Salt

Down in the Rancho

divided into about thirty compartments by means of temporary wooden bulkheads, and was stowed with enough sea salt to salt nine hundred or a thousand tons of cod. There was just room to shift the salt to have one compartment clear, and the deckboys had to shovel prodigiously to achieve this, throwing the heavy salt up from the depths of the hold and hauling it back clear from the hatchways. It was obvious that that salt was going to make a great deal of work before the voyage was over, for the whole lot of it would have to be shifted. There were about five hundred tons, and shifting it was considered primarily a deckboy's job. That he did not have to work salt was, perhaps, the doryman's only privilege.

While all this work was going on, the ship with her crowd of men had settled down splendidly. Of the ship's company of 70 men, 63 lived for'ard, 60 of them in the huge rancho with its three tiers of wooden bunks arranged like huge shelves, and the chief cook with his two assistants in a small cabin, which opened off the rancho on the starboard side. The cooks had the privilege of a cabin because their rest was so constantly interrupted by baking and by general galley duties. In addition to being cooks, the second and third acted also as the servants of the rancho, dishing up all the meals and cleaning up afterwards, and generally keeping the big house clean. This in itself I should have thought work enough for one of them; it was plain, even so early in the voyage, that nobody in a Banker had an easy time. The working-day was fifteen hours, and sleep was always a secondary consideration.

The cooks did well, though there was only one stove —and that with one coal fire—to prepare meals for seventy. There were two main meals a day: away from the fishing grounds, breakfast consisted only of coffee and rolls. It was good coffee, and they were excellent rolls. The rancho had its meals at 12 and 12.30, in two shifts, and at 5.30 and 6, again in two shifts. These meals were substantial, consisting always of stout soup, and a main fish or meat dish afterwards, with coffee or tea, and individually baked loaves of bread—all he wanted for each man. The fish was always well prepared, either boiled, baked whole with sauce, or fried in olive oil. Aft, we had a midday meal about 10.30 or 11 A.M., and the evening meal at 6 or 6.30. Meals were good and servings generous, but I had never learned to like anything prepared with or from bacalhau, though I had been eating the stuff at sea since I first went in Scandinavian sailing-ships, some 30 years earlier. We had a great many dishes prepared from bacalhau. We also had claret, cheese, and fresh fruit, for the owner had provided many cases of delectable Azorean pineapples, and the saloon deckhead was festooned with oranges and lemons hung in strings and arranged in patterns. The saloon was a simple hall in the after-part of the poop, almost filled by a square table and six swivel chairs, all carefully secured to the deck. A large radio receiver took up much of the after bulkhead, and this was generally kept at full blast on the wave-length allocated to the Portuguese Banks fishermen. A large steam radiator, a photograph in colour of the British Royal family, and a watch-bill, completed the saloon adornments.

The cabins opened off the saloon and off two short alleyways, one on either side of the engine-room bulkheads. Captain Adolfo had a large, comfortable cabin with a double-bed, and a bathroom adjoining, on the starboard side. Here there was also a radio-room, though telegraphy was not fitted, and in this was some modern navigation equipment including a barograph. The deck officers occupied the cabins on the starboard side, and the two engineers those on the port side. The engineer's boy lived in the rancho, and so did the cabin-boy. Everybody aft was from Ilhavo, except César: they were all friends. Captain Adolfo always addressed the young mates by their first names. They were both young enough to be his sons. They were a cheerful, happy lot, well accustomed to living together in ships, and I had no difficulty in settling down. But I had some anticipatory qualms about the life on the Banks.

Ten days out from the Azores, we were in fog and upon the Banks simultaneously. The fog lifted to show a fleet of large trawlers at work, and the Grand Banks lighted with so many fishing lights that the place was like a city street. These were French, Spanish, and Portuguese trawlers, and from the radio-telephoned conversations among them, we knew already that the fishing was poor and that the weather had been vile.

On hearing this, the Second Fisher, João de Oliveira from the Algarve, was depressed: with so many large trawlers scouring the Banks, he said, there would soon be no cod left, for the bottom was being spoiled, their feeding-habits ruined, and they had no chance. The mate, however, was not so gloomy. There would always be fish for the hand-liners, he said, for the dorymen did not fish where the trawlers were operating. There were 35,000 square miles of fishing grounds on the Grand Banks and only a fifth of that area could be trawled.

We were not able to fish anywhere at the moment,

because we had no bait, and we had to hurry into St. John's to buy some. We dribbled on with not wind enough to blow the fog away: and upon the Saturday afternoon the black bases of some horrible cliffs suddenly loomed in front of us, here and there whitened with ice and snow. Could this be Newfoundland? At first glimpse, the place looked more like South Victoria-land, in the grim Antarctic. But Newfoundland it was, as the bull-roars of the fog-sirens on Cape Spear and the entrance to St. John's harbour soon confirmed. As soon as it cleared enough, we went in under power to a berth at St. John's. It seemed a severe, rock-bound place, gloomy, depressed, and bitterly cold. We went in through a narrow entrance between two rocky hills, to find ourselves in a sort of crater lake, and through the heavy rain we could see half a dozen big Portuguese hand-liners moored alongside before us, including our sister-ships *Creoula* and *Santa Maria Manuela*, as well as the Viana do Castelo motor-ships.

"No bait," said the pilot. "It's a bad ice year—the herring aren't in yet; or if they are, it's beneath the ice. Bad ice-season, good fishing, they say," he added cheerfully.

No bait? This was a blow!

CHAPTER FOUR

INTERLUDE AT ST. JOHN'S

> And, having victual'd there their wearied Fleet,
> Proceed on their long course as it is meet.

WHEN Sir Humphrey Gilbert was at St. John's claiming Newfoundland for Queen Elizabeth, he had cause to note the "liberalities of the Portingalls" who then occupied the place, for the Portingalls "put aboard our provisions, which was wines, bread or rusk, fish wette and drie, sweet oyles, besides many other, as marmalades, figs, lymmons barrelled, and such like." The English, he says, were "supplied of our wants commodiously, as if we had been in a countrey or some citie populous, and plenty of all things."

It seemed to me in April 1950, three centuries or so later, that there had been some falling away: the present inhabitants could not supply the visiting "Portingalls" even with frozen herring to bait the hooks on their lines. Bait shortages were no new thing, nor were other difficulties about bait. Apparently, the descent of the greater part of the Portuguese hand-lining fleet upon St. John's so early in the season was a new thing, forced on the fleet by the absence of suitable bait-fish from their own coastal waters. The *Argus* had not visited St. John's before because she had always taken enough sardines at Lisbon to fill her refrigerated rooms and give bait for a month or two on the Grand Banks, and then she had gone into North Sydney, in Nova Scotia, for more bait for the Greenland season.

Now she was in a fix, and so were all the others. Nobody had any bait. Within a few days there were fifteen big Portuguese hand-liners lying at the St. John's wharves: before the week was out there were more than twenty. The motor-ship *Cova da Iria,* whose master fretted at the delay, went out to sea again to fish without any bait, and when I asked how he could do this, I was told that he would keep the long-lines aboard, and send the dorymen jigging. That is, they would fish all day standing upright in their dories, with a jig-line in each hand, flicking the jig, which was a barbed leaden fish, smartly through the bottom water in the hope of snaring hungry cod. This was a laborious method of fishing and not fruitful unless cod were abundant, and according to the trawlers, they were not. But it was better than idleness. One enterprising firm imported some frozen herring and squid from Halifax and Boston, but there was not much bait in those ports either, and the few remnants of Newfoundland's own banking fleet were also fitting out and demanding such bait as there was. However, a few of the Portuguese got to sea, using this old bait. The captains did not like old bait: they said that it softened quickly on the hooks, but it was better than no bait at all. They were afraid that it was the wrong time of the year for the cod to go for such bait as squid, but most of them agreed that the voracious cod would eat anything.

Baiting was not a mere matter of cutting up old fish and sticking the sections on the hooks. Captain Adolfo told me that the cod had certain clearly defined eating habits, though it was perfectly true that it would eat anything. At certain times of the year, when the herring were running into the bays along the coast to spawn, the cod became so excited chasing the schools of herring and filling itself on them it was apt to disregard all other bait. Later, it did the same thing with caplin, a small fish which came in great schools to certain parts of the Banks, notably round the Virgin Rocks. Later still, it used to turn to squid but, for some unknown reason, the squid had suddenly disappeared from the Banks and both the cod and the bait-seeking fishermen had to look for something else. Possibly the meandering of the Gulf Stream had caused the squid to go.

The herring, he said, should have arrived in the bays round Newfoundland by mid-April, and the cod should be there, following them in; but there was too much ice in the bays—for it was a bad ice season—and either the herring had not yet arrived, or could not be caught through the ice. The bait-fishers were accustomed to net herring in great quantities when it ran in to spawn, and it was this herring which the Portuguese were now seeking to buy. It simply was not there.

One of the first Portuguese words I learned was "paçiéncia"—patience. The schooner captains, who met in groups ashore to tramp from agents to consul and from consul to bait-trader's office—and all in vain—were forever repeating that word, to console one another.

Days grew into weeks. It was cold, damp, and trying. The smallness of the harbour compelled the considerable fleet of waiting ships to lie alongside, for which they were charged wharfage, in American dollars. Expenses mounted and they were earning nothing. The *Cova da Iria* was doing poorly, with her dorymen half frozen. The three-masted schooner *Santa Isabel* was caught outside in a vast field of pack which had drifted down on the Labrador Current, and it was reported that the last French hand-liner, the St. Malo-man *Lieutenant René Guillon,* was also caught in the ice. The nightly discussions of the day's trawling, which the trawlermen always carried on about the time of the evening meal, could be heard on the radio: there was never a note of optimism. Few fish, small fish—to a Banker the word fish means cod: there are no other fish to him—and atrocious weather. Ice, fog, gales. The ships which went out with old bait were doing badly, except the 70-dory motor-ship *Sernache,* which on a lucky day took more than 200 quintals.

Ah! Paciencia! Paciencia! said Captain Adolfo, and was echoed by his brother Almeida of the *Creoula,* and Captain Labrincha of the four-masted *Aviz,* and Captain Silvio of the *Elisabeth,* and Captain Matias of the big *Milena,* which had been an American schooner once, and was built in Florida in 1919. (Her timbers were enormous.) The four-master *Coimbra* came in with some damage, caused by the ice; and the ancient captain of the motor-ship *Capitão Ferreira,* an old ex-doryman of the hard-bitten school named Marques who was then on his fiftieth campaign, tramped the snow-filled, windy streets of St. John's so industriously in his endless quest for bait that he caught pneumonia, and had

INTERLUDE AT ST. JOHN'S

to be carried off to hospital—much against his will—so that when the bait finally began to come in, his young son David had to take the motor-ship to sea. David was his mate, aged 23, and the *Capitão Ferreira* was a pretty motor-ship of 750 tons, with a capacity for 12,000 quintals of fish in salt bulk, carrying fifty-three dorymen. Young David Marques was equal to the emergency, though the only other officer aboard was the second mate, aged 20, making his first trip to the Banks.

There were over a thousand Portuguese dorymen in the port, and their behaviour was exemplary. St. John's is a fairly small place: the town sits upon its waterfront and straggles up its waterside hills. To go ashore is to walk the main streets, for the whole place is sailor-town, and a most interesting sailor-town it is. But it is a city in which the seafarer is taken very much for granted: if there was a seamen's mission, I never saw it opened, and no Portuguese was invited to enter it. The thousand dorymen walked the streets, penniless (for they could have no money until the ships took fish, having already received advances for their families, before leaving Portugal) and amused themselves looking in the shops, which were well-filled with goods, and watching the passers-by. Whenever the weather gave them a chance, large groups of them went to the near-by streams, taking their laundry with them, and beat it clean on the rocks, for there was no spare fresh water in the schooners. Others kicked a cloth football about on the wharf, in which pastime they were joined by the south-side children. When school was over for the day, groups of little children came down to the wharves where the motor-ships and schooners lay, and skipped with the big Portuguese or swung on slack mooring-lines. It was a strange sight to see a dozen swarthy dorymen skipping and holding ropes for ten or twenty merry little boys and girls. The Portuguese were all family-men themselves, and they loved to play with the children. The children knew no Portuguese and the dorymen no English, but language was quite unnecessary. They understood one another very well and got on splendidly.

Our sailors had their dories, their oilskins, their lines, and everything else ready before we had been three days in St. John's, and we were there more than a fortnight. The ship was kept clean, watchmen were appointed, and there was plenty of leave. The dorymen were not bored, though restless at the delay: a spell of quiet rest was a new experience to them, and the older men enjoyed it. They had made their bunks in the rancho as comfortable and draught-proof as possible, and the steam-heat was working. They had their own heating system down there, and the boiler stood at one end, its smoke going up the hollow foremast and exhausting at the masthead. The main diesel exhausted through the hollow jigger-mast, so that sometimes the *Argus* presented the peculiar appearance of a large schooner under sail with smoke coming from both the fore and the jigger masthead. Peculiar appearance or not, the steam-heat was very welcome, and so, upon occasion, was the diesel, for to shift about the Banks always under sail would be most trying. That, however, I was to learn later.

Meantime, we lay on the far side of St. John's harbour, alongside the *Capitão Ferreira*, and we had put out a nest of dories to act as ferries across the harbour. I spent my time learning all I could of cod and cod-

INTERLUDE AT ST. JOHN'S

fishing, for St. John's was an excellent place for this, and I used the opportunity to visit other ships, and make the acquaintance of other captains. Our own group of particular friends—the captains of the *Creoula, Aviz*, and *Elisabeth*—were frequently aboard, always for at least one of the main meals of the day. Captain Adolfo kept a good table and the cook was excellent. These meals were merry, satisfying, and most interesting, except that the only subject ever discussed was cod. The captains frequently shouted at one another and gesticulated as they forced particular arguments home to one another, but since they all tried to force their arguments, statements, allegations, and indeed everything home in the same manner, I doubt whether much was really achieved by their vehemence. But it was all very friendly, and they loved it. Sometimes we all went to the *Elisabeth* for a meal, or to one of the other ships. There did not seem to be as much as a fish restaurant ashore in St. John's, though fish was the main industry of the place. If there was such a restaurant we never found it, or a fish market.

Meanwhile more and more of our fleet were taking old bait and going to sea. We could not do that because, apparently, our owner had contracted with another firm to supply the bait, and a rival had already bought all the old bait there was. At last our firm began to receive a little new bait, but the roads were so bad that the motor-lorries bringing it were much delayed. It was taken in a bay in the south, frozen, and then carried by road. Newfoundland, which had been much suppressed and exploited in its chequered history, bankrupt once and under commission government for some time, was extraordinarily backward in many ways, and outside the city limits the roads were mostly dirt-tracks, useless at that time of year. At the time of our visit, Canada had recently taken over the formerly independent dominion, at the request of a majority of the Newfoundlanders themselves: but there were many things besides roads to be put right.

Newfoundland had begun as a centre, ungoverned and ungovernable, where enterprising seamen and adventurers exchanged goods and fished. They chose the place for the very good reasons that it was near the Banks, and its lack of government made "free" trade possibilities unlimited and immensely profitable. It was a place to exploit, a place in which to make money. Early dictators forbade all kinds of settlement, in order deliberately to prevent colonisation and orderly development, for they feared that these would put an end to the unbridled individualism and great profits of their trade. Not all who sailed from Bristol for the Banks off Newfoundland were fishermen, by any means, though the men of Devon in Elizabeth's time knew the coasts of eastern Newfoundland as well as they knew the coasts of Cornwall and their own county. As early as 1518, Anthony Parkhurst, a Bristol merchant, writing to Hakluyt, said . . . "there were generally more than 100 sail of Spaniards taking cod . . . ; 50 sail of Portuguese; 150 sail of French and Bretons, mostly very small; but of English, only 50 sail." But the English "were commonly lords of the harbours where they fish. . . ." At that time, the annual fleets from Europe were manned by 15,000 men, and all were supposed to leave again when the season was over. Cod were always plentiful, but there were good and bad years for catching them, years of difficulties with bait, or ice, or weather.

INTERLUDE AT ST. JOHN'S

Newfoundland's own share of the fishing had declined greatly by 1950, though no English "lords of the harbours" had been coming, then, for at least a century. The Newfoundlanders themselves had been fishing with small schooners for many years. They were ideally placed for such fishing, and there was an assured market in Central and South America, and the West Indies. It ought not to have been beyond their business acumen and seafaring abilities to take and dry the cod which crowd in such great abundance along their coasts, more economically than ships and fishermen who came annually from Europe. European ships, moreover, had to accept the disability of heavy-salting the cod, which made it inferior to the lightly-salted fish prepared ashore. A deep-sea ship had no way to preserve her fish aboard other than by putting them into salt bulk, and pumping the pickle out daily as the salt absorbed the moisture, and so halted decomposition. Cod were more palatable if they were dried ashore, for the heavy-salted variety could become as tasteless as a sea-soaked board.

But the Newfoundlanders, or the exploiters of Newfoundland, failed, in the long run, to profit from their advantages. There were, in 1950, far more Portuguese Bankers in the port than there were Canadians, and great trawlers flying the flags of Spain, France, Portugal, and one or two of Italy, came in for stores, or diesel oil, or to land sick, almost daily. The only Newfoundland fishermen we saw were a handful of bald-headed diesel schooners lying at the wharves, each with eight or ten dories, and some of them for sale. A few small trawlers brought fresh fish to a plant by the harbour where it was filleted and fancy-packed for the American trade. The schooners were short-ended runabouts, spoiled for sailing, though their softwood hulls were shapely and seaworthy. They were large, for two-masters, except those for the Labrador fisheries. These fisheries were in such a bad state that there was serious talk of abandoning them altogether, at any rate for that season. The Labrador schooners were small, old and sometimes decrepit. They had no dories, for they acted simply as depot-ships for what was really bay-fishing. The cod running into Labrador fjords to spawn were taken in fish-traps, dried ashore, and brought back in the schooners, which took the fishermen and fish-driers along the Labrador coast each season. They were called "floaters."

All the Canadians I saw, both in Newfoundland and later in Nova Scotia, were schooners only by courtesy: they could no longer be properly sailed, and to have a serious engine defect was regarded as making them unfit for sea. The Portuguese, who had to make long voyages, had kept the sailing qualities of their ships; but a policy both shortsighted and unseamanlike had been allowed to prevail on the Canadian side. A "schooner" had riding sail, and that was all: if she *had* to sail, it was unfortunate. Yet many of these vessels could readily have combined the advantages both of sail and power, as the *Argus* and her sisters did. The Newfoundland fishing suffered from a long surfeit of feudal finance, and many of the fishermen, since their work so often failed (through no fault of theirs) to provide them with a living, had abandoned a calling so arduous and little rewarded. The island now had alternative employments, such as its paper-mills and iron-ore mines; and moreover, now that they were Canadians, the Newfoundlanders could seek better paid work elsewhere. No longer did Newfoundland build graceful wooden vessels to carry its fish to the West Indies and Brazil in the winter months, bringing back salt for the curing, and fishing in spring, summer, and autumn. The local Banking fleet had been allowed to decline from more than three hundred vessels shortly before the turn of the century to forty-odd in the early 1930's, and the total number of Newfoundland vessels over fifty tons engaged in Banks fishing in 1948—when there was still a considerable post-war demand for salt cod, and business was excellent—was thirty-nine. These had been reduced to less than a score by 1950. The Newfoundland coastal, Banks, and Labrador cod fisheries had yielded nearly 1,200,000 quintals of dry cod in 1947: a year later, with fish just as abundant, the take was less than a million quintals, of which less than 150,000 were caught on the Banks.

The highly skilled, but dangerous, profession of dorymen, of which the Newfoundlander had long been a courageous and extremely competent exponent, was in danger of decline, and already there was a marked reluctance among the younger men to accept its obvious risks and scant hope of reward. Poverty had been associated with fishermen for too long. Better distribution of fresh fish, largely through the spread of refrigeration, and the fierce competition from the well-packed agricultural products which could be prepared and marketed so much more easily, had dealt the salt cod trade some heavy blows. The younger people did not want salt cod, if they could have fresh-frozen fillets; many markets did not care for cod at all, if they could be supplied with tastier fish. Some good markets had been spoiled by undue exploitation, and by forcing them to take peculiar fish to which they were not accustomed during the war years, when food shortages, and government buying, had put the producer in a temporarily favourable position.

These problems were well understood in St. John's, and so was the risk of thinning the cod stocks seriously by the use of too many and too efficient trawlers, and a new kind of fishing recently introduced there by the Spaniards. While we were in the port a pair of very small Spanish vessels arrived and secured alongside astern of us. They were minute and identical, and their decks were filled with nets and cables.

"Ha," said Captain Adolfo, taking a moment off from his exhortations to patience, and looking coldly along the wharf. "*Parejas*. Who let them come over here?"

"Now the cod *do* run some risk of depletion," said Captain Silvio. "Those things have cleaned up many good banks off our coasts."

These "parejas," or pairs, were more feared than the trawlers, which had to keep to those areas of the Banks where the bottom could be trusted not to destroy their trawls, which were expensive. The Spanish pairs fished by throwing great nets round an area of water, and then drawing the nets and catching every living thing. If they were introduced on the Grand Banks in force they certainly would do the hand-liners no good.

The Spaniards had many trawlers, as well as these pairs. The trawlers were manned, in the main, by young men. Spanish berets, Portuguese stocking caps, Breton and Basque fishermen's clothing were to be seen along Water Street, St. John's, almost any day, for there was generally at least one French trawler in port from the extensive fleet of more than two score of them outside.

INTERLUDE AT ST. JOHN'S

These came into port only when they had to, and wasted little time there, but the Spaniards, apparently, had some agreement with their crews by which they were given spells ashore at regular intervals. A fine big Italian trawler called the *Genepesca IV*, with French technicians aboard, was in for some time. She had a tripod mast supporting a queer sort of gallows for her net-handling equipment, and she looked a smart and seaworthy vessel. There were, all told, upwards of a hundred big trawlers working the Banks, forty-four of them French, and another forty Spanish and Portuguese.

More interesting than the trawlers were the Arctic whalers and sealers, though the whalers were not operating. We were there at the time when the sealers were coming back from their brief spring runs. Two secured astern of us. They were crowded little vessels, reeking with the pelts of murdered seals. One was a converted war-built wooden tug, fitted with radar and the latest navigational aids, and with steel sheathing at her bow. The ice had ripped this sheathing from her and torn away part of her bow. Cutting her cruise short, she returned to port with the torn bow "protected" by some beaten-out oil-drums. For two days, this vessel lay at the wharf discharging bloody seal skins with the blubber on them, and a great deal of the poor brutes' flesh, which waterside ghouls stripped from the flayed hides as they lay on the quay, and skewered on bits of old wire. Later I saw them carrying the flesh away, as well as seal flippers, also skewered on old wire. Groups of sealers stood at the street corners offering the stuff for sale. Another sealer came in with 17,000 pelts, and according to her crew, they had killed and maimed several thousand other seals which they did not take.

A few days later, two Norwegians came in, one of them the well-known *Norsel* which had already been to the Antarctic that season, to land the members of a scientific expedition. The *Norsel* was towing a compatriot whose propeller had been ripped off in the ice. The Norsemen at least tried to take aboard all the seals they killed. They were skilled seamen who followed the sea steadily, not hooligans with guns recruited from the waterfront streets and crowded into a seal-slaying sea-slum for a few bloody weeks, let loose upon the defenceless seals without, apparently, even first demonstrating that they could shoot straight and hit what they were aiming at.

At last we received our bait. By then, half a dozen ships had sailed—all the Viana do Castelo contingent, and several large motor-ships, and the *Creoula* a day or two before us. It was the second of May when our herrings came—60,000 lbs. of them, enough to fill both refrigerated rooms, for we had to take some for the schooner *Hortense,* and the *Creoula* took some for the *Gazela.* We worked all night taking in the bait, for already we had lost much time and all the advantages we had hoped to gain by sailing early.

On the morning of Wednesday, May 3rd, we slipped quietly from the St. John's waterfront. The harbour was fog-filled and the stony hills covered with snow almost to the water's edge. The pilot said we had come in the wrong time of the year: it might be spring in June, and there was generally something of summer in July.

Ugh! The harbour froze us to the bone, and outside the sea was fretful and the clammy sky wept, while the

INTERLUDE AT ST. JOHN'S

ship jumped and rolled. We passed the *Cova da Iria* coming back from jigging to collect her share of bait, and we headed towards the southwest, to a good spot on the Banks. By mid-afternoon the fog had cleared a little but the sea was up, and the sails would not steady her. At the wheel in the first night watch I heard the First Fisher speaking of plenty of fish on the morrow, and he grinned delightedly. His watchmates, gathered in the darkness round him (for they always kept their watch in a space clear of dories, near the wheel) were quietly singing some song of the south. There we were, at last upon the eve of the real thing, and on the morrow they would trust their lives to frail dories—on the morrow and a hundred other morrows—in frail dories, against the murderous North Atlantic, at its coldest and most relentless, on fretful shallow banks notorious for fog and sudden storm.

And now they sang.

CHAPTER FIVE

ON THE GRAND BANKS

> They did invade the Sea with Saile and Oare:
> Actions so proud, so daring, so immense.

I HAD been told, in Lisbon and at St. John's, that there were elaborate rituals connected with the first fishing, and with the first selection of a suitable anchorage on the Banks. According to my informants ashore, there were all sorts of signs which indicated the whereabouts of large stocks of fat cod, waiting for the hook—signs on the sea, in the sky, even mysterious indications known only to the codhunters in command, veterans of the Banks such as our Captain Adolfo. But Captain Adolfo simply went straight to a spot where he thought there was a reasonable prospect of finding cod, took a sounding, measured the temperature of the sea, anchored, and got on with the job. There was no ritual, and there was no time wasted.

It was the morning after we left St. John's and the weather was still bad. As soon as it was daylight, the bait was distributed. This was done with an elaborate care which hardly seemed necessary. Each man had a cardboard slip on which was the number of his dory. These slips were kept in the charthouse, and given out each morning. Then the man went along for'ard to the hatch by the refrigerated chambers, and joined the queue. As his turn came, he handed in his slip, the mate carefully counted him out a dozen frozen herrings, which deckboys had hauled up in baskets, and crossed off that doryman's name on a big card. Off went the doryman at once to cut his herrings and bait his hooks. I could understand the care with which he did that. The deck was soon crowded with men preparing bait as quickly as they could, for they knew they would not have time enough to bait all their hooks. They could only prepare the bait while there was nothing else to do, and as soon as all had received their bait, the dories were launched.

"Why be so particular about the bait?" I asked the mate.

"Aha," he said. "Some men are bad fishers—they use too much bait and they bring back too few fish. You see, we cannot carry bait enough to supply all the hooks the dorymen use daily, and they shoot those long-lines three and sometimes four times a day. So bait has to be guarded very carefully, and we have to see the men don't come back in the queue twice. They all want more bait. They all *need* more bait. We haven't got it. They all have to get some bait for themselves."

"How will they do that?"

"From cod, from the rocks, from little birds. We give every man a small line for catching birds. He catches them with hooks baited with liver."

"What, *birds*? Do cod eat birds, too?"

"Cod eat anything they can get at," the mate said. "But these are special birds. There are none about now. I will show you them when we see them. There are plenty in Greenland."

Bait from cod, from rocks, from little birds? Surely, cod were bottom-feeders. There was more in this codhunting business than I had realised. Later, I often heard Captain Adolfo discussing this bait question with his brother captains over the radio telephone: this was one of the subjects he discussed every day, and sometimes many times a day.

Meanwhile, the dorymen drew their bait in silence. A green doryman dropped a herring and another shouted at him. It was considered bad luck to drop bait before putting the dories out. Obviously, bait was something to be treated with respect. It was true that it had cost the ship a good many dollars and had already put her to a lot of trouble.

It was noon when we anchored. The position was 44 54 North, 49 29 West—about a day's good sail from St. John's, towards the south-south-east. There were no other ships in sight, and no fish either. There was still a high sea, but little wind, and the *Argus* pitched and rolled. I should have hesitated to launch a lifeboat unless it had been really necessary; but now they were going to put more than fifty dories out into that sea. After thirty years in ships, this was a new kind of seafaring to me. Until now, the sea had always been a means of supporting the ship and, by the skill of her people in their manipulation of the sails and use of the ocean winds, of moving cargoes about the world. Away from port, you stayed in the ship. Your world was bound by the vessel's rails and rigging, or at any rate you fervently hoped so. And now here was a vessel which cheerfully put out an anchor upon banks in the open sea, miles from the sight of any land and with no possible shelter (an anchorage to me had always been at the very least a partly-sheltered roadstead) and then proceeded to send away her crew in fragile little boats, without as much as a life-jacket between them or an air-tight tank in any of the boats.

It was wretchedly cold, and raining. The cold sea swept across the schooner's low decks. The one-man dories looked neither particularly strong, nor especially seaworthy for the work they had to do. They were already laden with tubs of baited lines, buckets of bait, light masts and oiled sails, oars, bailers, anchors and sisal lines, and each held its occupant's personal container, made of wood, brightly painted, and holding his tobacco, a flask of cold water and a small loaf of bread, a few olives and perhaps a piece or two of Portuguese ham sausage, and a watch, if he were rich enough to own one. This was generally tied up carefully in old newspaper, where it kept company with a whistle, or a conch to blow in fog. Knives, boat compass, sinkers, jiggers, a honing-stone, and grapnels for the long-line, completed the dory's cargo. When the doryman himself jumped in, in his heavy woollen clothing and oilskins and giant leather boots with wooden soles an inch or more thick, the dory had quite a load before it began to look for fish.

Out dories! was the order, and out they went with a will and a rush. There was a tot of brandy for each man who wanted it—many did not—and the brandy was dispensed with more generosity than the bait. The cabin-boy poured out the brandy from a kettle. It was traditional for the ship to offer a tot to each doryman before he set out on the day's fishing, just as it was traditional that life-saving gear should never be used. If a man's dory could not preserve his life, nothing could.

There was competition to see which side would get all its dories launched first, the Azoreans or the Mainlanders. The ship was rolling so much, her four yellow masts swinging like pendulums against the grey, wet sky, that I feared some of the dories must be smashed. But these men were experts at getting boats away, and a perfect drill had been worked out, probably centuries earlier. The schooner's low sides and low freeboard were a help. The dories were plucked from the nests by overhead hooks which fitted into the protruding grommets spliced into bow and stern. These hooks were manipulated by simple tackles led aloft, with the hauling part by the rail, so that a man or two on each tackle could swing a dory easily from the inboard nests to the rail. Here its doryman hurriedly adjusted its thwarts, saw that the plug was in, climbed the rail and jumped in himself. As the ship rolled towards that side, the tackles were let go at just the right moment and down went the dory with a rush and a thwack upon the sea. Immediately the iron hooks disengaged themselves, alert mariners hauled them back inboard, the doryman shoved off from the side for his life, and dropped astern. Once clear of the ship's side, his little dory seeming now smaller than ever, and dancing and leaping in the sea, he rigged his mast and little oiled sail, and away he skimmed towards the horizon, to choose a place to lay out his lines. In a few moments, the grey sea all round was covered with the little dories, skimming away like dinghies jockeying for the start at some stormy regatta. But this was no regatta.

I was amazed at the good seamanship of it all, even when I reflected that this kind of thing had been done by the forebears of these very men—and many a good man from Bristol and from Devon and Cornwall, Ireland and Wales, Brittany and Normandy, and Massachusetts, Newfoundland, and Nova Scotia—for more than four centuries. Not an accident, not a shout, not a swamped dory nor a lost piece of gear. In twenty minutes, we had fifty-one fishermen out upon the sea, and the bare decks of the *Argus* looked strange. There were less than half a dozen dories left. I had imagined that the dorymen would not go far from the ship, but many of them crossed the horizon and were out of sight.

A little after five o'clock that afternoon, the recall signal was made. This was a large flag, made from a Tate and Lyle sugar-bag, hoisted at half-mast in the jigger rigging. It was nearly seven before the last dory was back. There was scarcely a dozen cod among the lot, not anything like enough to begin salting. The cook took the entire catch and looked for more. Captain Adolfo, with a shrug, was weighing the anchor long before the last dory was back, seeking a fresh place for the morrow's fishing, and when all the dories were nested again we stood towards the south-east, under power. Power was used to hoist the patent anchor from thirty-five fathoms down, and I was told that the usual way of shifting about the Banks—of which a great deal was done—was with the motor. By that means, the men could get on with cleaning and salting, and any spare time they had could be devoted to preparing bait and lines. The sails were used for big shifts and long passages, and sometimes on the Banks. Some schooners had to use their sails much more, even for small shifts of position, because they had weak motors.

The dories were hoisted inboard in the same manner in which they had been launched, by means of the iron hooks and the tackles, which remained permanently in position. The most that anyone caught the first day was two cod. Even the First Fisher caught nothing. I heard Captain Adolfo passing on the sad tidings to various friends, by telephone. The *Creoula,* it seemed, had done little better. Captain Silvio, in the *Elisabeth,* was just coming out of St. John's to try his luck.

"That man will get fish," César said. "He always does."

"Why?" I asked. "How can he know where the fish are?"

"He doesn't know," said César. "But the fish know where he is, and they go there."

If this were true, I hoped we should see the *Elisabeth* very shortly, and fish the same water, taking good care never to let her out of our sight. But this, I gathered, was among the things not done. We should certainly be seeing a great deal of the *Elisabeth* and of the *Creoula* and probably the *Aviz* as well, for these were our particular friends. But Captain Adolfo was quite capable of finding places to fish without assistance.

And that he proved, the very next day. It was again a grey, cold day, with a sigh of wet wind in the rigging and the water bitterly cold. It was the kind of day when even the dawn was unwelcome, for it served only to emphasize the misery. At the first crack of light—a little after four—the dories were hoisted out, and I reflected upon the great amount of four-o'clock-in-the-morning courage which was needed, and taken for granted, in this sort of seafaring. The shelterless dories looked like invitations to frost-bite, and the bleak sea was restless, and heaving, and contorting itself without rhythm and without cessation, as it frequently does on shallow banks. Away skimmed the dories across the grey face of the sea, and within a few moments, less than a dozen of our fifty could be seen. We had left a doryman in hospital at St. John's, another had fallen down the hold, and a third was sick. Throughout the day, we could never count more than perhaps eighteen dories from the ship, even from aloft, with powerful binoculars. I hoped they were all right, but I was the only person aboard who worried about them. The two mates and the few deckboys left aboard were busy clearing the decks, and rigging electric cluster-lights along both sides of the fore-deck, the better to illumine the night working. They also prepared a series of small troughs, from wood and canvas, down which they told me the cod's livers would be washed by running water towards the steam oil-pressers for'ard.

The recall went up at five. Within fifteen minutes the first dory was back, laden to the gunwales. Every few minutes, the doryman had to bail furiously to maintain himself afloat. This, however, did not displease him nor did it displease Captain Adolfo, who permitted himself a grin (after first sighting about a dozen more dories in much the same state) while he gave orders for the anchor to be weighed at once. When the anchor was off the bottom, we drifted down to leeward, picking up dories as we went. This was done to save the heavily-laden dories from having to work to windward, which was dangerous and might have swamped them. In the days before power windlasses, many dorymen had been lost that way. The very first thing which was given power in the Bankers was, indeed, the windlass. The old-timers had to anchor with a great scope of cable and many heavy rope cables to remain safely at anchor at all; and the very use of these things made their working extremely difficult. It used to take them anything up to eight or ten hours (and even more) to recover their anchors: the *Argus* had hers tripped in two minutes, and left it out at the end of a couple of shackles, until such time as she would anchor again.

For the next two hours, we were picking up our dories. They came alongside for'ard, first, the Azoreans to port and the Mainlanders to starboard, and Captain Adolfo sheered the ship this way and that with the trys'l, to give the more heavily laden dories a lee. The First Fisher's dory was laden so that it had no perceptible freeboard at all, aft, and the First Fisher of St. Michael's and the Second Fisher's were in much the same condition. Captain Adolfo, in brightly painted yellow clogs, with a fur-lined leather jacket flung across his shoulders, wet his pencil as he entered his estimate of each man's catch in the little black book he kept for this purpose. He looked in the dories, moved into such lee as he could find, and wrote down his estimate of how much dried cod the fresh fish in the dory would yield, measured in kilograms. The fish were not weighed or measured in any way, but no doryman gaffed up any cod until he had seen the dark countenance of the captain peering down over the rail at his dory. A cold wind came up, from the north, with the sting of ice in it: nobody worried about it, though the

ON THE GRAND BANKS

water was lapping at the counters and along the low sides of half the dories. As each came alongside and its catch was recorded, the doryman was handed down a gaff by a deckboy, and he began at once to heave his fish up and over the rail, into the pounds. This in itself was a labour of difficulty, for the cod were big, and the schooner's sides high to a man in a dory. The roll of the ship sent the dories jerking in all directions, though they were supposed to be held in position by deckboys. There were often four dories discharging along each side, and a dozen others waiting off to take their places. As each dory was discharged it was moved quickly along the ship's side, aft, where it was at once hoisted inboard, stripped quickly of its thwarts and bulkheads, and nested. As each doryman came aboard, he joined the hoisting teams, without any time off for a hot drink or a bite to eat, though there was coffee waiting in the galley all day. No one ever came for it. I saw a few men take drinks of cold water. No one ate until all the dorymen were back, the fish all in the pounds, and the dories nested and lashed down. The dull cod, rubbery-lipped and cow-eyed, fell with heavy plops into the pounds, where they slid about with the vessel's rolling. As night came, the cluster-lights were lit, and the whole foredeck was brilliantly illuminated. When the dories were all back, we anchored again, and the men had a hurried but substantial meal. The pounds were almost full with considerably more than two hundred quintals of fine fat cod—more than forty ton of fresh fish.

As soon as they had eaten, all hands fell upon the job of cleaning and salting, for there would be no sleep for anyone until that was done. The work was well organised, and each man knew his place and what was expected of him. Each doryman was graded according to his skill as a specialist in some part or other of the cod-cleaning and salting process. About twenty climbed down into the hold, and these were the salters and their helpers. They worked under a Salt Foreman, a quiet Algarvian more than fifty years of age who had been fishing since he was a boy. Thirty or more formed themselves into gangs on deck. The main gangs consisted of three men, one standing in the pound, where he picked up the cod and, with three smart strokes of a sharp knife, opened the gut and all but severed the head. He was called the "throater," and was considered the least skilled. His three cuts made, he slapped the wet cod smartly upon the working table by the cleaning-vat, where a second man with one movement of a gloved hand swept away all the innards. Then this man skilfully removed the liver (easily distinguishable), and pushed it through a hole in the working table to the liver chute immediately beneath, while at the same time he was breaking the head from the fish. Heads of small fish he threw overboard at once, but all large heads he pitched into a corner of the pound where a deckboy, looking very miserable in the cold, cut out the tongues and saved them.

A third doryman, the most skilled member of the team, took the headless and gutless fish from the "liverer," and with one blow of a big square-bladed knife, split it so that it lay open to the tail, on its back. Then he cut the upper part of the backbone out, with two further strokes, and pushed the split and cleaned cod into the vat of running sea-water to be washed. He was the "splitter," and that was the team—throater, liverer, splitter. The deckboy was a "tonguer," for the moment;

he would have further duties in the morning when the dorymen had gone, for then he would have to fall to on the backbones, and remove whatever was edible from them. A green doryman pushed the trolley-car, coming to each vat in turn and gaffing out the cleaned cod. When it was full, he pushed the trolley to the salt-chute, released the inboard side (which was hinged) and out poured the fish to slide into the hold. Here the salters and the salt passers and gaffers took charge, the gaffers working on their knees flinging the fish down to the salters, who were crouched along a tarpaulin at the bottom of the hold. They shouted "Fish! Fish!" as they gaffed the fish down, and the salters had to beware lest a large wet cod lay them flat. They were as expert at dodging cod as they were at salting them.

Under the watchful eye of the Salt Foreman, and using salt from tubs which the salt-passers kept constantly replenished for them, they scattered salt along the fish and stowed them on their backs, fore-and-aft, with the utmost speed and dexterity. From time to time, Captain Adolfo came to keep an eye on this part of the proceedings, which was vital. The key-men in the fish-cleaning were the salters and the splitters. A splitter could soon spoil a fish worth perhaps a pound, with one false stroke of his knife: unless they did their work perfectly, the salters could spoil the whole cargo.

Every man—throater, liverer, gaffer, splitter, salter, salt-passer, trolley-pusher, tongue-boy—worked magnificently and without pause. Nobody took a moment off for as much as a cup of coffee, though they must have needed the stimulant, and there was plenty in the galley. For'ard a gang, led by the second mate and the assistant engineer, appearing ogre-like in the glare from their boiler, toiled to convert the pale livers into oil; a liver boy hustled about saving livers which had missed the chute. The mate, in a Greenland cap, was general superintendent of the deck, under the watchful Captain Adolfo who was never far away. The tall boatswain from Ilhavo stood at the head of the Mainlanders' line, and a fine old man with a gentle face at the head of the Azoreans. I watched this man. He was a really magnificent splitter. He looked like a bishop and he split great cod at the rate of ten or twelve a minute, while the pound where he worked diminished at a rate perceptibly faster than that of any of the others. His name was Jacinto Martins; he was fifty-nine years old, veteran of thirty-nine voyages, and the father of four lovely girls at Ponta Delgada. Three of these were married: he hoped that soon the fourth might marry, too, for she was nineteen. Then perhaps he would need to come to the cold Banks no more.

The cold north wind increased steadily, and the snowflakes swirled in the strong light of the cluster-lamps. The wind began to sigh and howl in the steel rigging, and the sea water, breaking over her as she rolled, washed bloody and gut-filled about the decks. In the big pounds, the dead cod slid and slipped about almost as if they were alive. The snow and the water everywhere glistened on the oilskins and the strong, unshaven faces of the dorymen, who had begun that day at four A.M. The First Fisher, splitting fish and ripping out backbones, gaunt and purposeful, reached a bloody hand into his sou'wester to pick up a cigarette. Each man to his own job, silent, effective, and energetic, making every knife stroke count, ankle-deep in water and often knee-deep in cod, cods' heads, cods' guts—

Waiting for Bait at St. John's—Some of the Viana do Castello Ships

A Meal in the *Argus*, at St. John's

Doryman with "Grippers" to Protect His Hands

The Dory Is to the Doryman What the Horse Is to the Cowboy

ON THE GRAND BANKS

hour after hour they worked on. Nine o'clock, ten o'clock, eleven o'clock; nobody struck the bells, and they had forgotten the loudspeaker. Soon after eleven it was done, except the backbones, which could be finished by the deckboys in the morning. They could go below. A generous meal was served, of fish soup. The Soup of Sorrow, they call it.

"Eat it not," said César, "for he who eats will come to the Banks again."

I heeded him, and ate it not, for it was made from cods' heads and halibut faces; and I did not want to come to the Banks again. This was going to continue for six weeks or so, and we should then move on to Greenland, where conditions were much worse. I thought that one voyage would be enough for me.

When the work was done, some of the dorymen began to clear their lines which had been caught on rocks during the day. The *Argus* was fishing on a rocky bottom, at a place where trawlers could not interfere, and some lines were snarled and others lost. They had to be clear by the morning, and no time was allowed for this. A doryman fished at least a twelve-hour day, then cleaned fish while there were fish to be cleaned, or salted them. The rest of the time he slept, ate, kept a night-watch (which was done in pairs), fixed his lines. He was lucky if he averaged four hours' sleep in the twenty-four, but sometimes a spell of bad weather gave him a longer break. This, however, was irksome to him, for the ship was not taking any fish. He was not earning, nor was the day advanced when he could return to his home and family.

Over the soup of sorrow, Captain Adolfo regaled us with talk of a campaign when he was able to work the

Argus for fifty-three days without a break—fifty-three days of fishable weather—and all hands were so tired that they used to fall asleep into their soup: but they filled with 14,000 quintals, and were on the way back to Lisbon well before the end of August.

This was the kind of voyage they all liked: but there seemed scant prospect of it in 1950. It would be a month and more before we could go to Greenland. Meanwhile, the day after that good fishing, bad weather kept the dories aboard, and the following day the catch was about twelve quintals. The fish had gone. But where?

We weighed, and shifted again, and tried once more. And then weighed, and shifted again, and tried again . . . and tried and tried and tried. And sometimes took good fish, but more often not. And sometimes rode out icy gales, and watched the sleet lie in the dories and in the stowed sails, while the ship lay to her storm trysail on a long scope of cable. In the fogs, we cried with our great siren, lest the steamers run us down, and if the dories were out, sounded our distinctive peal on the old church bell, for them to hearken.

Days grew to weeks, and still we had not fished the first thousand quintals. *Paciência!*

CHAPTER SIX

BELLS IN THE FOG

> To tell thee all the dangers of the Deep,
> (which Humane judgment cannot comprehend)
> Suddain and fearfull Storms the Ayre that sweep:
> Lightnings that with the Ayre the Fire doe blend;
> Black hurracans; thick nights; Thunders that Keep
> The World alarmed, and threaten the last End:
> Would be too tedious, indeed vain and mad,
> Though a brasse Tongue and Iron Lungs I had . . .

D-O-N-G! Ding-ding-ding! D-o-n-g-d-o-n-g-d-o-n-g!

The mate struck the *Argus'* distinguishing signal on the big church bell, which hung always ready from a frame secured in the starboard mizzen rigging. Our siren was wailing, and so were the sirens of half a dozen other Portuguese hand-liners we could not see. The different note of a Nova Scotian schooner could be distinguished, clearly, and the mate had thought he heard one of our dorymen blow our signal on a conch shell. The big fog sirens of the air-raid warning type all sounded the same: a doryman seeking to return to his ship and coming close to her, steering on his compass bearing, sometimes makes the distinctive signal with a whistle or a conch, that his ship may answer. He knows then that his course is right. Ships hearing dorymen making signals other than their own do not reply, and the big church bells were kept, apart from emergency, for their own lost dorymen.

Day after day their tolling sounded gloomily through the Grand Banks fog. It was fog when the dorymen were called, at four o'clock in the morning; fog when they had finished their twenty-minutes breakfast and preparations; fog when they drew the day's bait, when they hoisted out the dories, and slipped quietly off into the endless wet murk. It was fog all day, and all night. Fog did not deter the dorymen, but it had an adverse effect upon the fishing. Though a few intrepid spirits did not allow the densest fog to prevent them from sailing for miles, the majority kept nearer to the ship on foggy days, lest they became lost. Being lost temporarily was commonplace and accepted, but to be lost hopelessly was dreaded. Dories from the Banks had been found, with their lost occupants, after many days, and dories had been rowed and sailed to Canadian ports. There were stories that at least one had drifted right across the North Atlantic and, like the lumber-laden derelict schooner *W. L. White* which was abandoned east of the Delawares in the middle of March, 1888, fetched up after months among the Outer Hebrides. But I could verify none of these stories, though it was true that at least one schooner, in the old days, had been driven as far as the Azores when her cable parted, and an American two-man dory had made a special crossing of the North Atlantic.* It was also true that more than two hundred dorymen had once been lost in a night.

A dory might be able to cross the North Atlantic but

* In 1867, from Gloucester (Mass.) to the English Channel, for a wager.

a doryman adrift without adequate preparations could not, and survive. None was anxious to make the attempt. But none ever took precautions against such an emergency. On the foggiest days the dories were prepared in precisely the same manner as on clear days. No doryman took any more water, or more food, or as much as a hurricane lamp to light at night. Only a few plutocrats had flasks to keep some coffee hot. Though they were out of sight of the ship half a minute after they were launched, none ever hung back or looked perturbed in any way, about anything. I marvelled at their fortitude for, after all, they had only small boat compasses, and often they were out of sight of the ship so rapidly that they could not even get a useful bearing. On these occasions, each would go off along his own course and try to keep to it; but they sailed, if they could, and the flat-bottomed boats made considerable leeway. And what of the Labrador Current? A compass bearing from the thwarts of a small dory could be poor guidance over five or six miles of the treacherous Grand Banks.

If I were a doryman, I thought, I would keep very close to the ship when the fog was dense. A few did so, but very few. There was a belief among them that they would never find fish in quantity close to the ship, though centuries ago all fishing was done from the ships themselves, and cod were plentiful enough then. Whether there were fish near the ship or not, nearly all the dorymen preferred to go off at least a mile or two before shooting their lines. In fog and wind alike, on the whole the Algarvians were the most fearless dorymen. One old Azorean, a tiny man scarcely four feet six inches high and thin in keeping, with a wizened face, a stubble of grey beard, and a look of perpetual mild astonishment in his watery blue eyes, never went very far from the ship, whether it was fog or clear. Whenever there was a slight clearing, his unfilled dory would at once loom into view. It was number twenty-two.

"Ah, you spiritless bait-waster!" Captain Adolfo would moan. "Be off with you and fish!"

But he would say nothing to the small man himself. A doryman was master in his own vessel, at any rate in fogs: captains rarely interfered, or gave direct orders as to where dorymen were to go or how to fish at any time, and then only when absolutely necessary.

The little man almost never brought back a fair doryload of fish. He was about sixty years old, a native of Horta, a doryman of nearly half a century's experience. Once he had been good: now they called him the Little King because he did nothing, which was a libel on kings. But the men liked to have him about. He was a clown and a jester down in the rancho. So he shuffled off in his dory every morning with the rest of them, and he rowed a little way and then stopped; and if the weather was bad, he put out only a little of his longline and kept the rest aboard, because he wanted to be able to recover his line quickly if a squall came, in order that he should not be impeded when the recall was hoisted. The Little King was always first back, whenever the recall went up. In fogs, the recall was a signal on the big siren, one very long blast followed by the ship's distinctive signal.

The Little King had nine sons. He did not really need to fish at all. So he did not care greatly how many fish he took: he was still a doryman because he knew no other life. Going to the campaign for half the year had for so long been part of his existence that now he could not stop.

The *Argus* suffered the clownish Little King, cheerfully, but her other dorymen were fishers of a different calibre. She could afford only one clown. The indefatigable First Fisher filled his dory day after day, even when no one else could half-fill, until I began to think that what was said of him might be true, that he had a private store of large fat cod somewhere on the Banks. The First Fisher of the Azores, a cheerful, handsome fellow named Francisco Martins; the Second and Third Fishers from those islands; the Second, Third, Fourth, Fifth, and Sixth Fishers from the Algarve, always fished extremely well. I soon observed that about twenty of the men caught half the fish almost every day: most of the others had to find fish in abundance before they could fill a dory. The measure of a good fisher was his worth on bad days: anyone could fill a dory when the fish were there. Captain Adolfo put up a list of the men's fishing results each week, giving his pessimistic estimate of the value of their catch in kilograms of saltbulk cod. Not that it made much difference. Captain Adolfo was an ultra-pessimist as a catch-estimator, but the men would be paid, at the end of the voyage, upon what the ship landed, each sharing in proportion to the catch credited him. His weekly estimate was therefore intended more as an indication of the doryman's share than as an accurate record of the catch, though the captain always under-estimated by a fixed percentage. He knew what he was doing and he had to be a pessimist. One thing was certain: the ship would never discharge more fish than she had loaded. Cod shrink.

On the lists and in the dories, the First Fisher was

consistently ahead of everyone else. The First Fisher of the Azores was Third Fisher of the *Argus:* almost all the leading fishers were from the Algarve. The Fuzeta-men were more experienced both with long-lines and with dories than the Azoreans or the men from Ilhavo —we had a few dorymen from there—because they used similar fishing-gear at home. The others did not. The Azoreans fished mainly by other methods in their own waters, and the men from Ilhavo rarely fished at home. In Ilhavo it was considered undignified for a doryman to fish through the winter in home waters. He got himself a job as a rigger or a rigger's mate, either in his own ship or in one of the schooners in the harbour at Gafanha, or he might, if he were really hard-up, take a job in the salt-pans. If possible, he preferred to do nothing at all. Few attained this enviable state. There were no bachelors.

The dangers of the doryman's life are traditional and remain unchanged—the dangers of swamping, of being overwhelmed in the sea, lost in fog, run down by shipping, smashed against his own ship in a seaway. Alone upon the North Atlantic, often out of sight of his mother-ship and a hundred miles and more from the nearest land, his boat a little thing of fragile planks without power of any kind and without even a rudder, with a home-made sail and an oar or two his only means of making headway through the water, compelled to over-load if he is to fish usefully at all, and forced to remain habitually upon the turbulent water of the fogbound Banks or the stormy shallow seas off Greenland, never knowing when he leaves his ship in the morning that he will come safely back again at night, his food in the dory always cold, his dory with-

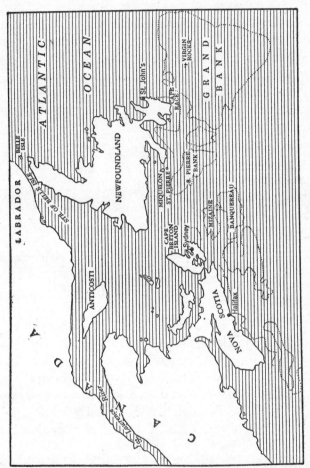

The Fishing Banks Off Newfoundland and Nova Scotia

out shelter and as exposed as a raft or a piece of driftwood, it is certain that if he were not sustained by his religion and his long traditions, he could never venture to be a doryman. If fishing by dory were a new industry, there would be no dorymen.

Yet there were other disabilities and dangers of which I had never heard, and the first was sea-sickness. For the first few days, some of the green dorymen and even one or two of the seasoned men were seasick in their dories as they pulled away from the ship's side. When I first sailed in a dory, the motion appalled and almost sickened me. That you can be seasick only once in a voyage and, after that one attack, remain well, is a landsman's belief. As the great albatross of the Roaring Forties, which soars five thousand miles in glorious health and wondrous freedom, immediately becomes seasick when caught and kept upon a ship's deck, so may an experienced seaman be rapidly sickened by a marked difference of motion, and the difference between the decks of a 700-ton schooner and a one-man dory is marked. The schooner is almost quiet, compared with the dory. The jumping of a dory in the open sea is unending, relentless, and without rhythm. Only in a calm is it quiet, and there were few calms in April and May. To shoot a 400 to 600-hook long-line from a leaping dory, while cold, wet, and violently seasick, is a veritable test of endurance. I did not see any dorymen slacken his efforts through seasickness.

Then there were whales. I had not heard of these as a menace to dorymen, and when a family of humpbacks played round the ship for several days I did not appreciate the wariness with which the dorymen regarded them. They were a bull, a couple of cows, and a baby about eight or nine feet long, a merry little fellow who spent most of his time under his mother's flipper but frequently left that shelter and, coming to the surface on his own account, surveyed the ship with an impertinent eye as he stood on his tail, almost vertically in the water. We could clearly see the baby clasped in its mother's flipper, generally on the left-hand side, as the family came to the surface. They were close enough for us to see the wart-like knobs on their noses. One morning in mid-May, when about eight of us were in company and the sea was littered with dories, the humpback family came on the scene again: it was not until the evening that we knew that one of our Azoreans had had a dangerous brush with the largest of them.

While he was quietly hauling his long-line about ten o'clock in the morning, our Francisco de Sousa Damaso of dory 57—a green fisher from Ponta Delgada—suddenly saw the bulk of a whale rising beneath him in the clear sea. The whale's blubbery bulk was already tangled with his long-line, a circumstance which the whale had probably not noticed, though a surprised cod or two were flapping against his barnacled blubber. He failed to notice the dory, either. He came up to blow almost right under it, toppled it until it all but capsized, and some of the gear fell out. Francisco Damaso, no stranger to whales, hurriedly cut his line and prayed. The dory slid off or the whale slid away and, for a split second, an amazed doryman glared at an astonished whale. Then the whale blew, sounded lazily, and departed, still with the long-line trailing astern of it and Senhor Damaso's morning take of cod going off to oblivion. Thankful that he had not been spilled into the sea, the doryman sculled about and picked up his gear, which was floating; then he recovered all the line he could, and continued fishing. A good doryman, green or otherwise, does not return to his ship before his dory is filled, or the recall has long been hoisted. It was after six o'clock when he came back and asked for some more lines.

Our Azorean was very fortunate, for the same day the whale had spilled another man into the sea, from one of the *Creoula's* dories. A dory from the schooner *Groenlandia* had also been swamped. Fortunately, both dorymen were rescued, though their boats were lost.

The same week, the captain of the *Dom Deniz* (a shapely three-master from the Monica yard near Aveiro) was astonished to see, in a clearing from the almost perpetual fog, one of his dories drifting silently by his ship, empty. He knew the dory, for most ships painted their names on all their dories, and those which did not used other distinguishing marks. He knew moreover that its occupant had been gone only a few moments, for the long-line was still hanging from the bow, where the man had been hauling it in. So the captain of the *Dom Deniz* immediately leapt into a dory himself and pulled off to rescue his man, if this were possible. By the grace of God it was. He found the man, a strong swimmer, treading water not far from the ship; and before long, the man and the dory were at work again. It appeared that the doryman from the *Dom Deniz* had been spilled out of his dory by a slow flick of the whale's tail. Until he was thrown out, he had not seen the whale.

That was three brushes with a whale, perhaps the same whale, in the one week. I heard of no other such encounters throughout the remainder of the season, and all the captains said such incidents were rare.

We knew at once about the happenings in the *Creoula* and the *Dom Deniz* because the captains spoke of them in the evening yarnings and exchanges of news by radio. On foggy days, there was a greater use of the radio telephone. When the *Dom Deniz* had the man in the sea, her mate at once alerted all other ships to watch for him, while the captain went in the boat. We never missed anything on the radio. When he was on deck, Captain Adolfo had the loudspeaker there connected and switched on at a discordant full blast: it was on equally loudly and discordantly in the saloon all day and all night. Even when he was down in the hold supervising the salting, a hatch was kept open and the dark face of our captain would appear there frequently, an ear cocked to the loudspeaker. The only time he was not listening was when he was using the microphone himself, but he was no microphone-hog, as a few of the captains were. There was one especially who chattered half the day. He was something of a wit though he had a caustic tongue, and his brother captains bore with him. After all, they had several wavelengths available to them, if the talkative captain monopolised one. Moreover the life they all led was exceedingly monotonous, and very trying: they would put up with much for the sake of a little diversion.

Nor was the loudspeaker for ever blaring the pessimistic discourse of the codhunters, for whom no Bank had fish enough and all days away from home were far too long. Once a week, there was a special programme broadcast from Portugal for the men on the Banks, in the trawlers as well as in the hand-liners. A recording unit went round to each of the principal fishing-ports in turn, and recorded brief messages from wives and children at home. On Sunday evenings, these were broadcast to the ships. These programmes were awaited eagerly and for them the loudspeakers achieved new triumphs in volume; while the dorymen, the deckboys and the officers, the cooks and the engineers, listened with almost painful eagerness to hear word from their loved ones. It was not what their people said that mattered; it was just hearing their voices.

Flaps of sou'westers were pushed back from hairy ears, bright eyes were turned towards the metal speaker, as if to peer to some sunny street in Portugal three thousand miles away, and to catch a glimpse of wife and children. Even a splitter's knife would pause in mid-air for a second as he heard his name mentioned by the radio announcer, and a great smile on his whiskered face would instantly disperse the appearance of an austere and relentless automaton it had borne an instant before. These programmes were called the *Hora de Saudade*, the hour of longing: * they were the only relaxation the dorymen knew, and they were tremendously appreciated. They were arranged by the Gremio at Lisbon, and have been a feature of the Portuguese Banks fishing for more than a decade.

I liked the way the Portuguese did things, and the more I learned of them the more I liked them. Here was a sailing-ship, a schooner in the old tradition, a vessel designed to be moved by the wind. Magellan

---

* The precise translation of the Portuguese word "saudade" has defeated many abler linguists than myself. At least one book has been written about its meaning. The English word "longing" is not at all adequate, but there is no better.

could have stepped aboard, or Queiros, or Corte Real, Cabral, da Gama—any of them—and within the hour, have mastered the clarity of her simple rig and aids to navigation so that he might have sailed her about the world, or round the world, were he so minded. Any of their crews could have come aboard and felt at home almost upon the instant. Yet she made use of every modern aid which was of value either navigationally, or for safety, or spiritually. A man who had never been in a sailing-ship in his life could equally well come over the rail, and—if he were a seaman at all—be at home aboard in a day or two. Her radio telephone, her loudspeaker system, her refrigeration, her echo-sounding, air-raid sirens, electric light, cod-liver oil plant, the factory arrangement for the bulk handling of fish, her steam-heat and hollow masts—all were of use, directly, and their value was apparent. An ancient Banker, veteran from some tiny lugger which used to come banking with a fire-box on her salt and what she caught to cook, might stand agape for a moment and marvel at a thing or two. But he, too, would be of use to her the moment he cared to turn-to, and he would find aboard a great tradition which he could understand. Nor would he find much changed in the nests of dories, if he knew anything about such craft. If he did not—and there have been dories on the Banks only during the last century —he would probably be as appalled at the idea of using them as any modern steamship mariner might be.

Our *Argus* was old, and new; ancient, and modern. She was a part of life, and very like to living. Even the loudspeaker's blaring against the moaning of the rising wind was not incongruous. The electric light, the big bait rooms, the patent anchors, the powered windlass and the big diesel down below—these things were Godsends, and no hoary ancient mariner would deride them. Deride them? Good God, no! Thank heaven for them; for the life was hard enough still. The more facilities she had for working, the greater cargo must the ship win, to pay for them; the more aids they had and the better their long-lines, the harder must the dorymen work, too, and the greater the risks they must run. In the old days a Banker would be manned roughly on the basis of a man to every forty or fifty quintals of fish needed to fill her, for this was a fair average for a good man to take in a season. But in the *Argus*, each man had to take more than 200 quintals, and bear considerable stress of mind until her great capacity was filled. Her owners must suffer a heavy loss if she was not reasonably filled, and the livelihood of seventy families could be jeopardised. Her long-lines with their scores of thousands of hooks required vast quantities of bait, and the bait was used up whether it took fish or not. If it did not take sufficient fish, then the money invested in the refrigeration and the long-lines was wasted.

I thought of all these things as I watched the splitters and the throaters and liverers at their work, and listened to the cries of the liver-boy, and the chatter of the gulls squabbling astern over the tremendous feast provided for them. There was plenty of time to think, time in the fogs, and in the gales; and every day. I often wondered what the dorymen thought, in their dories, or the deckboys, as they got on with the prodigious amount of work which was their lot every day. The dorymen did not think about survival, unless the subject was forced upon them. They took that for granted. I asked the First Fisher more than once what came into his mind during the long hours he was at work in his dory. Work, he said, mostly work. His mind was on his work, and the weather. He watched the behaviour of the sea, for the sets affected his hooks; he watched the surface for signs of the small fish on which the great cod thrived, for he knew that if he saw such fish, he must alter his own methods. He watched his gear, which must be kept perfect. As he hauled his long-line, he watched every hook, and again as he rebaited and paid out at the end of the hauling. As he fished with his hand-lines, while the long-line was down, he had to keep his brain in his fingertips, he said, to know where his jiggers were, and so the better snare the cod. That was his own expression—his brain in his fingertips. He had a wonderful touch with a hand-line. Even at forty fathoms, he knew at once when his jigger was near the bottom. Some of the men scarcely knew at all: it is not easy, in a current or in a tideway. But he admitted that sometimes there was a moment to think about home.

As for the deckboys, whose day was endless toil, who never knew the satisfaction of bringing back a full dory, who daily scrubbed the litter of cods' innards, scraped backbones, and tongueless heads from the big decks, and cleaned all the fishing gear—the vats, the cleaning tables, the salters' tarpaulins, the gaffers' chutes—and worked salt by the score of tons, cleaned tongues by the barrel-full, scraped membranes skilfully from big cods' backbones by the thousand, as for the deckboys, poor devils, they scarcely had time to think. The ambition of all of them was to become dorymen, and that as soon as possible; for at least a doryman did not work salt, and he was in command of his own dory, and of his soul while he was with her. Some dorymen were younger than some deckboys. These young men were from Algarve, protégés, sometimes relatives, of the older men, who had brought them. Most of the deckboys were graduates from the Gremio's Fishermen's School, or were from Ilhavo.

Perhaps, sometimes in the fogs, the deckboys were glad they were not in dories, or when the sea suddenly grew high and vicious, and the dories had to come back in great danger. On such occasions the men had to jump for their lives, to leap over the rail at the right moment when their fish had been gaffed up. Sometimes as the schooner rolled her rails under, a dory would almost be washed right aboard. This had happened, and indeed there was a story of a captain who had once collected his fifty dories in a sudden gale when it was hopeless to wait for fish to be gaffed from them, by running down to them and sheering his schooner alongside each dory in turn so that they rolled inboard as the seas washed over the schooner, each with his take of fish intact. I heard this story, but I did not meet this captain, or hear him named. Such a thing might perhaps be done, but hardly without damage. We did not care to let the dories wash aboard for fear that they would be smashed. Unless their recovery were orderly there could easily be bad accidents, and loss of life. On several occasions dorymen had to swim for their lives, as a result of bad judgment or bad luck: by the grace of God, Captain Adolfo always picked them up. They were always pitched out beside the ship.

One reason why the captains liked the schooner-hull, with its long main deck, was because it did make it possible to sail the dories right aboard in a seaway. It also made normal recovery of dories a much simpler process than in the motor-ships. These had light decks built up above the main-deck before the bridge, where the dories were stacked seven high, six and eight nests abreast. To fill such a space with so many dories, when there was room to hoist only one a side at a time, took longer than enough in emergencies, though some of the very latest motor-ships had electric winches to assist. A doryman in a big motor-ship had more chance of accident than his counterpart in a schooner, and the captains were well aware of this. The dorymen did not like the big motor-ships, though not for this reason. They did not like them because they said they were too hard to fill, and so they always stayed on the Banks a long time. Though in theory the motor-ships required precisely the same average fishing to give them full cargoes as any of the schooners, and less than the *Argus*, in actual fact their size seemed to discourage the dorymen. The mere knowledge that the ship required 16,000 quintals to fill, and 16,000 quintals was a prodigious quantity of fish, weighed upon their minds; and being discouraged, they did not fish so well. A doryman's spirit was very obviously of first importance in the success of line-fishing. A 12,000-quintal ship was big enough for him, and a 14,000-quintal ship was too large for comfort. A 12,000 ship which could carry 14,000 if she took them easily was perhaps the ideal. The *Argus* was in this class, and most of the motor-ships were not.

By the middle of May, the best day's fishing we had was two hundred and fifty quintals. (Captain Adolfo said it was two hundred and thirty; but Captain Silvio, passing by in his motor-ship and having a quick look, said it was three hundred.) The fish filled all the pounds and spilled across the decks, until the whole working platform was jammed with them. Everyone had a full dory that day, except the Little King. Even he had filled half his dory. The First Fisher filled his twice, until it scarcely had freeboard enough to remain afloat. That night the scuppers ran with fishguts and blood until one o'clock in the morning, and the oil presses could not cope with all the livers pouring down the chutes. This was good fishing, but there were no smiles. A really good day, Captain Adolfo said, was one when he took five hundred quintals, and he had done that only once in thirty years of Banks campaigning. The day he did so—it was on the Greenland grounds—the *Creoula* took six hundred, and he thought that was the record catch for any hand-liner. The fish reached up to the booms and filled the whole fore-deck, and cleaning and salting had continued until eight or nine next morning. After four hours' sleep, the dorymen had gone off again, and they took more than three hundred quintals on the second day. This had been during the '39-'45 war, when none but the Portuguese and the Eskimos fished the Greenland grounds.

In 1950, we had no 500-quintal days: if their telephoned reports were to be believed neither did any of the other captains. But their broadcast statements were not always to be accepted: we learned later that the *Soto-Maior* had taken five hundred quintals one day. She had seventy dorymen, but this was a good effort none the less.

The day we took two hundred and fifty quintals there were fifteen Portuguese hand-liners in company—too many ships, Captain Adolfo said, as he shifted twenty miles in the dark hours. Why not remain where the fish obviously were, I asked. They would not be there on the morrow, he said, but I was at a loss to understand how he could know such a thing. He was quite right. The ships which did not move took poor catches the following day. Unfortunately, so did we. Where did the infernal fish get to? The captain explained that May was a poor month in a bad ice-season, because the cod had not settled down. The herring were late this year, and this probably upset the cods' normal eating habits. The cod were grubbing about all over the Banks instead of congregating in the spots where good feeding was. The trawlers, with their constant dragging of heavy trawls across the sea-bed, had spoiled some of the feeding-grounds, and now no man knew precisely what was going on. All the captains knew places where the cod should be, where the bottom feeding was good and the trawlers could not go; but to be upon just the right place, with fishable weather, just when the cod were there too—that was the problem. When the cod moved in in strength, they soon cleaned up an area. They grubbed up every crustacean and scoffed every living thing, including the young of their own kind, if any such were foolish enough to be about.

There were crustaceans, some lance, and young herring in the stomachs of most of the cod we took. I could not see how the wretched cod could get any joy from its life, grubbing about in cold waters mouthing mussels from rocks forty fathoms down, growing fat for a few years by its predatory and cannibal activities, then to be caught upon a Portuguese hook or swept into a heedless trawl, suffocated, and decapitated. Surely the cod's must be one of the dullest lives in all the sea, for it never even fights.

It was difficult to have any sympathy with the foolish cod, and I prayed for a few 500-quintal days as fervently as any doryman. I thought of the cod as consumer of our bait, a filler of dories, and a cargo for our hold. The masses of them lying in our pounds excited no sympathy, only a desire to see the pounds better filled with more of them, fatter and larger.

I prayed for some 500-quintal days, but we did not get them.

CHAPTER SEVEN

THE PROBLEM OF BAIT

> Thence, many nights and many sadder days,
> Betwixt rough Storm and languid Calms we grope.

WE OFTEN saw Canadian schooners on the Banks. They were quite different from the Portuguese, and easily recognised. They were invariably painted black, with long, low hulls, two stumpy masts stepped close together, the fore setting a small gaff-and-boom foresail and the main always with a rag of a trysail, triangular in shape, with a very high foot and little more area than a fair-sized handkerchief. None had a bow-sprit, and none had topmasts. There was usually at least one in sight, and often there were three or four, but we did not see more than this. At times there were more than a dozen Portuguese in sight, most of them big schooners.

The Portuguese were painted white because white is the best color to see in fog. The captains, and even the dorymen, had had to be convinced of this fact, but it was so. The black Canadians were harder to see. They all fished with two-men dories, and they came close enough for us to see how they worked. Their dories looked very much larger than ours, and each man had a pair of oars. They seemed to stay much nearer the ship than the Portuguese, though frequently we noticed that their ship stayed with them, slowly under way with her engine turning over and her scraps of steadying sail holding her. They hauled their long-lines four times a day and left them buoyed at night, and sometimes in the daylight hours as well. When they hauled in, one man could be seen taking the cod off the hooks, while the other re-baited these hooks and paid the line out again. There were usually not more than eight or ten dories at work from any of them. When they had hauled their lines and shot them again, they pulled back to their schooner, or she ran down and picked them up, and they gaffed the fish aboard, hoisted in the dories, and immediately got on with cleaning and salting.

In this way they appeared to fish more steadily than was possible under the individualist methods maintained in the *Argus* and her consorts. But Captain Adolfo said their methods required prodigious quantities of bait, and the Canadian schooners could not remain at sea for more than a few weeks at a time. Their hope was to catch a lot of fish quickly, and if they did not, as was frequently the case, their bait was gone, their endurance finished, and they had to go back to port for more bait, fuel, and food. They paid all hands by share of the catch, and this was not the Portuguese tradition.

There certainly were advantages in the two-man

THE PROBLEM OF BAIT

dory, Captain Adolfo admitted—for he did not take the view of the hidebound traditionalist—in that they were safer, more seaworthy, and could carry more fish. There were advantages in the two-man hauling and baiting team, too, and in the idea of leaving the long-lines down to catch fish while the crew cleaned the previous take. Each country had its own methods. The French, when they had hand-liners, used two-men dories, but they set them round the ship in a clockwise pattern which was contrived to fish the Banks methodically, and to catch all the fish in the area. This had its advantages, but the long and the short of it was that if the fish were there, all ships filled, French, Canadians, Portuguese alike. The Portuguese, Captain Adolfo said, had experimented with the two-man dory, which the Gremio would like to see used generally for the sake of its greater safety. But the dorymen were against it, and it had failed. The schooner *Oliveirense* had been fitted with two-men dories this season, and was supposed to be using them. Captain Adolfo appeared to have considerable doubts whether she was using them or not. The *Oliveirense*, a beautiful three-master, built of wood by Monica's, was a vessel we had not seen on the Banks, though she had been near us at the Blessing.

Both in St. John's and later in North Sydney, I learned that, whether his methods were superior or not, the Canadian doryman had certainly not been averaging anything like the catch made by the Portuguese. It was a poor Portuguese doryman who did not take 150 quintals, each of sixty kg. of fish, even in a bad season. The average catch by a Newfoundlander, I was told, was nearer forty quintals, each of only fifty kg. He made four trips to the Banks in a season, but he did not go to the Greenland grounds. If he could not fill from the nearer Banks, he did not fill at all. He was sometimes in trouble with the trawlers, which ploughed through his buoyed lines by night and in fog, destroying his gear.

This was a trouble the *Argus* experienced but little, for her lines were never out unless the dories were with them, and she was gone from the Grand Banks before the trawlers and the Spanish "pairs" were there in full force. Like all the Portuguese, she did her best to keep out of the trawlers' way. One passed near us, trawling, once only: we wished her no harm but it was with a little quiet satisfaction that we watched her stop to repair nets, torn on the rocky bottom. There was room enough for us all, and there was plenty of good bottom where she could go dragging. Much of the Canadians' frequently expressed opposition to trawlers was based upon the traditional schooner-man's dislike for seeing his investment made obsolete. This did not apply in the Portuguese fleet, for many of the largest trawlers were owned co-operatively by the schoonermen themselves. To some extent, the same owners controlled trawlers and hand-liners, and each type of fisherman had an interest in the welfare of the other. Moreover, the captains of most of the trawlers were Ilhavo men, like the captains of the hand-liners, and the crews were recruited from the same centres. There was, apparently, little interchange of crew between the two types.

Our dorymen, far from looking enviously at the trawlers' crews who never had to go overside but simply cleaned and salted what the nets dragged in for them, abhorred the idea of going in a trawler, for two

THE PROBLEM OF BAIT

reasons. One was their natural pride in their own skill and the satisfaction they derived from being handworkers. The other, and perhaps the stronger reason, was that the trawlers tried to make two Banks campaigns each year and were therefore gone from home for at least ten and a half months in every twelve, while the hand-liners tried to make one annual voyage suffice. The investment in the trawlers was much greater than in the hand-lining ships, even the most modern of them, and the only way an adequate return could be earned was by fishing for as much of the year as possible.

The French trawlers tried to make three voyages a year, if they could. To do this, to fill those big ships with cod three times, they had to range far and wide, scouring the Arctic seas and all the cold waters they could find, and sometimes take very small cod. We could hear them sometimes, on our radio, talking about cod prospects and results in places as far away as Spitzbergen, Iceland, the Barents Sea, and off the coast of Labrador. They talked of these places as if they were Londoners quietly discussing the railroad services to London's suburbs, and when the Banks fishing remained bad for several weeks without sign of improvement, we heard several French captains announce their intention of going off to the Barents Sea, as if the place were twenty miles away. This they did, but the hand-liners and the Portuguese trawlers remained where they were.

"Grand Banks first, then Greenland when the ice allows us," Captain Adolfo said. "If we start chasing round the North Atlantic and all the Arctic looking for cod we shall never find them, and spend all our time looking. Our programme is fixed."

Our programme might be fixed but, for the time being, our cargo was growing with painful slowness. The weather throughout May continued bad, and we were doing well to launch the dories three days a week. Often, even then, they had to be recalled before they had shot their long-lines more than once. In one week our total catch was less than 200 quintals: our best week on the Banks was less than 800. It had been hoped to take at least 3000 quintals on the Grand Banks before going on to Greenland, and the early sailing from Lisbon was designed to achieve this. In 1949, the *Argus* had taken less than 2000 quintals off Newfoundland and, even after fishing the Grand Banks twice, had returned to Portugal in mid-October with less than a capacity cargo. The dread of having to do this again was very real and constant. Everyone hated the thought of having to move down from Davis Straits to the Grand Banks for the second time, and that in late September and October, when the winds were violent and the weather even worse than it was in spring. The Grand Banks in the spring, off Greenland in the summer and early autumn (with as little of autumn there as possible)—that was the hope, and that programme should fill any hand-liner. But it often did not.

Our bait stocks were steadily declining. We had taken bait at St. John's only for the Grand Banks fishing, with the intention of replenishing at North Sydney for the Greenland campaign. Captain Adolfo complained that some of the dorymen were using ten herrings to catch one fair-sized cod (possibly the Little King's average actually was about a herring to a cod), and the bait-fish in the refrigerated chambers disappeared at an alarming rate. To provide bait for all the dorymen took about

THE PROBLEM OF BAIT

20,000 slices of frozen herring, and each herring gave only twelve or fifteen pieces worth putting on the hook. The balance of about 10,000 hooks was baited with pieces of cod-roe, with crustaceans taken from cods' stomachs or torn from the rocks by hooks and snoods, and with other usable odds and ends retrieved at the nightly fish-cleaning.

Providing expensive bait for 20,000 hooks three times a day took bait in vast quantities, and the trouble was that the bait went whether it caught fish or not. Once removed from the refrigerated room and impaled on the hook, a piece of herring did not keep for long. If, as often happened, the doryman was able to get it in the water once only while it was fresh, its chances of catching anything were correspondingly limited. On days when bad weather forced a late start, the dorymen were issued with herring enough only to bait the long-lines for two hauls, or one, depending on the time remaining that day. When they were recalled early for the same reason, herring not already on the hooks was placed in the refrigerators again for use on the morrow. Captain Adolfo, and everyone else, watched the bait like a hawk. It was more important, by far, than fresh water, of which there was an ample supply, because the *Argus* was a steel ship. (Wooden ships could not carry so much and were often short. For this reason, and because of the high initial cost of the equipment, some of them did not carry cod-liver oil plants.)

One of the severe disadvantages of bad weather—especially in fits and starts—was the waste of bait it caused. If the lines were not baited they could not be used: unless the long-lines were down a reasonable length of time when they were baited, they were wasted too. Yet again and again the recall had to go up before the dorymen had had time to make even one useful haul. A codhunting captain shows his skill by his ability to conserve his bait supplies: a foolish captain could easily use up all his bait when fish were not plentiful, and have no bait when they were. In the old days several dorymen used to be kept catching bait and doing nothing else, and each ship did all she could to replenish her own supplies. This she did by putting down baskets, baited with old meat, cods' heads, and the like, which took a sort of snail in great quantities; by catching all the squid she could when the squid were running; and by following the caplin in towards the coast and scooping them up by the barrel-full and salting them, whenever there was an opportunity. But the trawlers had spoiled the bottom where the snails used to be taken; the squid had gone—no one knew why—and none of the big hand-liners could hope to take bait enough herself to supply the prodigious needs of anything from fifty to seventy 600-hook (or even 400-hook) long-lines a day.

The old-timers had not used these long-lines, but had filled their dories with hand-lines and jiggers. A hundred-ton or two hundred-ton lugger could be filled that way, but it would be quite hopeless to try to jig up a cargo for a fifty-doryman, or to fill her with a cargo of cod taken solely with hand-lines. A few ships were still without refrigeration for their bait, and these continued to use the older methods. But they did not go to Greenland and none of them was larger than 200 tons. Neither did they manage to fill every season, or anything like every season; none of the nine which had remained on the Grand Banks in the 1949 season, for instance, had

THE PROBLEM OF BAIT

managed to fish full. Some of them were only half full, even by October.

As for the squid, these used to be found moving in vast quantities about the Banks, coming and going mysteriously. The general practice was that the doryman on night watch kept a line with a jigger over the side, and from time to time, as the spirit moved him, did a little jigging to see if he struck squid. If he did have the good fortune to find this succulent bait striking his jig hooks, he immediately ran to the Rancho companion shouting *"Lula! Lula!"* which is the Portuguese word for squid. Up came the rest of the dorymen at once and they all jigged away so long as the squid were there. A member of the night watch who shouted "Lula!" when no squid were jigged was far from popular. A great deal of sleep used to be lost, trying to catch squid. All in all, I don't know that it wasn't a good idea that they had gone. The cod had to eat something else.

The few ships which had taken old squid from St. John's for their Banks fishing were reported to be doing well, so the cod had not lost his taste for that delicacy. Our dorymen begged some frozen squid from friends in one or two of these vessels, but they ate them themselves. They cut them up, threw the pieces casually on the top of the galley-stove—the cook never seemed to mind, so long as a decent quiet was maintained while his pet chicken slept—and then plucked them off and ate them with relish. Out on the Banks, there was some exchange of news as well as squid, among the dorymen, especially when a crowd of ships was together. The afterguard had their news by the radio telephone, but the dorymen had theirs by word of mouth, and it was frequently more reliable. The whole of the time, we had a very good idea what the other ships were doing, sometimes, too good an idea, and Captain Adolfo occasionally had moods of slight despondency at the news that such-and-such a ship had fished over 200 quintals when we had taken sixty, or when word came from her dorymen that the *Condestavel* was already well past the 2000-quintal mark when we had not taken 1500.

"Maybe they are small quintals," I said, hoping to relieve the gloom which was heavy that day, the fifteenth consecutive day when our take had been rated at less than a hundred quintals.

Captain Adolfo refused to be comforted. His general attitude, and that of most of the captains, was one of quiet confirmed pessimism. He was a kind of an optimistic pessimist. He hoped for the best, expected the worst, and accepted what the Lord gave to him. His character was by no means gloomy, but the fishing made him appear so, rather often. Not that we endured a life of gloom. Far from it. Meals were always cheerful though cod was our staple. We ate cod fried and boiled (but always fried in olive oil); we ate cod fillets, whole cod, cod steaks, cods' hearts, cods' tongues (fried in batter), cods' cheeks boiled and fried (and these were prime favourites); dried cod boiled, shredded, minced, made into fish-cakes; we dried our own cod, and ate them in a dozen different ways; we ate some membraneous stuff scraped by the deckboys from cods' backbones, called Samos, and served with a boiling of potatoes and tomato sauce. We ate cod in every guise, but we always ate it, though sometimes Captain Adolfo, who had been eating it for more than thirty years, became so tired of it that he ate some bait for a change.

The Fish Were Washed in Vats of Running Water

Fish Cleaners at Work

Full Pounds—a Good Day's Fishing

Dories Back, Good Weather

The Mate Sounding the Fog Bell

The Happy Doryman

Sometimes Dorymen Had to Swim for Their Lives—Here a Dory Has Swamped and Overturned, Alongside the *Argus*

An Azorean Veteran

He Had Made 35 Voyages

The Brothers Salvador and Estrela Martins, from the Algarve

THE PROBLEM OF BAIT

We ate halibut, too, and haddock, which were fairly plentiful.

A typical lunch would consist of soup, followed by boiled cod with potatoes, raw onion, and scraped garlic for those who wanted it; then cods' tongues fried in batter; and then cheese, and coffee, with rolls of excellent quality, and very good claret. There was no limit to the quantity of the claret we might drink but no one ever had more than two glasses. We did not always eat two fish courses at the same meal. Generally there was a meat dish of some kind—salt meat (which Captain Adolfo abominated almost as much as he did boiled cod), or ham, or canned meat. On Sundays we frequently had canned Portuguese tunny, or sardines, and canned greens. We always had vitamin pills, after the eleventh day at sea—small red pills which rolled round the plates until they were captured, and were swept down with a mouthful of wine.

The food was good, the company excellent, the talk congenial though almost always of cod. Captain Adolfo was far older than any of the other officers. César the engineer was the only other officer more than 30 years of age. The Captain spoke to them all by their christian names. The two boy-mates he had known at Ilhavo since their childhood, and their parents before them. The assistant engineer had begun his sea-career at the age of fifteen, as cabin-boy with our Captain when he had command of the schooner *Creoula*. Our assistant-engineer, an industrious and competent young man, was the son of a boatswain from Ilhavo who had lost his life on a Banks voyage. The young man was studying electrical engineering through a correspondence course in Spanish, which was sent him from some school in Los Angeles. When he was doing nothing else, he was learning English. Our mate and our second mate had been scarcely a dogwatch at sea, by the old standards. They were handsome, likeable young men who worked hard, spoke little, and rested less. They had learned some of their seafaring in the academy, some in the merchant service, and the rest in Bankers. I liked the way the officers sat down to their meals at the common table. There was no false "side" about any of them, and Captain Adolfo never thought it necessary to erect barriers, the better to maintain his dignity and the eminence of his position. He had no need of any such artificialities. There was a sort of happy democracy in the schooner's afterguard, and her discipline suffered nothing for that.

The chief distinction between Captain Adolfo and his mates was that the captain carried almost all the burden of the ship, and worked even harder than they did. I could not grudge him the delight he derived from the infernal radio, for ever blaring in our ears. His dark countenance would light up at some jest of the volatile Captain Silvio's, and he would chuckle over his plate of fried cod cheeks. Or he would stop suddenly, and glare at the offensive loudspeaker above the saloon table when some other captain, fancying himself as a radio comedian, indulged in the low tricks of that calling at the expense of his brother-captains, or perhaps uttered some more than usually monstrous untruth about his take of fish. They all spoke untruths about their fish. It was part of the game, but there were some rules.

Captain Adolfo played the game as well as any of them. After each day's fishing, he would talk to his

THE PROBLEM OF BAIT

particular cronies about our luck, or lack of it: he always under-estimated our catch, bad as it might be—and over the radio it was never good—by at least forty per cent. Sometimes his brother in the *Creoula* would shout back at him (for Captain Almeida, in our sister-schooner, usually under-estimated only by thirty per cent) and there would be a great battle of words between the brothers with Captain Silvio joining in and the whole fleet enjoying the fun. No one enjoyed these occasions more than the brothers themselves. The microphone was a boon to all the captains—so much so that I wondered how they had ever got along without it.

The radio, of course, had more serious uses. On bad days when there was a faint hope of better weather, Captain Adolfo would call his friends and the captains of all the hand-liners he knew to be in the neighborhood, to discuss the weather they were having, to help in that vital decision which only the captains could make—whether to launch the dories or not. Each captain had at least fifty men's lives in his keeping. They hated to lose men. They hated to endanger men. But how could they fish, unless the dories were launched? The problem was always the same—was the bad weather getting worse, or might it soon get better? If it could be trusted to improve, then some measure of risk to the dories was perhaps acceptable. If it were worsening, the dories must stay aboard. The dorymen always accepted unquestioningly any decision that was made.

These decisions were difficult, and the captains sweated over them. Most of the captains affected a certain appearance of toughness where their dorymen were concerned. Few ever praised their dorymen, **individ-**

ually or collectively in public. Several of the captains had been dorymen themselves: they had all grown up in a hard, old school. No one had praised them. Praise and soft words were not part of their seafaring traditions. But I soon saw that Captain Adolfo was acting a part. He was the hard-boiled shipmaster on deck, and a harsh expression of critical disapproval rarely left his face. But he had to remember to put it there. Once he had a foot on the after companion it disappeared at once. He was in fact very proud of most of his dorymen, and zealous for their welfare at all times. Most of them he knew by their first names, which he used frequently (when the fishing was good). Some of them had been fishing with him for more than a quarter of a century, in schooner after schooner.

So he was anxious that their lives should not be unduly risked, and he and all the other captains went to infinite pains to spare and to preserve them. Sometimes they failed. The *Creoula* had already lost one doryman that season. This, all hoped, was sacrifice enough, and more than enough. The *Creoula*, in her time, had lost a number of men—once five men at the same time—but the *Argus* had been more fortunate. In twelve years she had lost only one doryman. The *Argus* was one of those ships which spared men, but the safe return of her large crew to port, year after year and season after season, with a full or at least a profitable cargo of Arctic and Grand Banks cod, had been achieved at the expense of much grief on the part of her captains. Captain Adolfo had commanded her during eight of her twelve voyages. It was a good record, and he did his best to keep it.

Yet the dories must go, and the fogs must imperil

THE PROBLEM OF BAIT

them, and the storms arise. Reluctant to send the dorymen out when the weather was bad, Captain Adolfo never hesitated to recall them immediately it was necessary or seemed to be necessary, despite the inevitable loss of bait and time. He never left the deck until all were safely back. Again and again, I stood with him as the dories came; sometimes when they had been able to fish only an hour or so, sometimes in a strength of wind that would have had us shortening down even in a great Cape Horner; sometimes in cold and silent fog, wet and weeping, perhaps the greatest enemy of all. He was proud of his dorymen, proud of their skill in finding the ship, and of their expert seamanship as each dropped warily alongside, into the broken water there which was often more dangerous than the sea itself (though Captain Adolfo always did what he could to quiet it). I remember a day on the Grand Banks when, after mid-morning radio consultations with a dozen ships, our captains and the others decided to launch dories in a stiff nor'wester which looked like easing, only to find the wind rising to alarming force within the hour, and the sea with it. The North Atlantic can be a treacherous ocean. It was dangerous work, getting the dories back. The ship was rolling and plunging, with few people aboard until the dorymen returned. The dorymen had to be most circumspect in the manner they approached, lest they be stove or swamped —swamped by the rush of waters or the breaking sea, or the water gushing from the washports as the schooner rolled. Frequently her rail went right under, scooping up the sea, and all this water gushed through the washports again, imperilling any dory unfortunate enough to get in its way. There were many washports, some near the pounds. The dories must approach them.

When each dory had been emptied of fish, its occupant leapt nimbly over the rail. The dory must then be nested down. The thwarts and bulkheads were rapidly rigged down once it was inboard, the tub of line hoisted out, with its bait; then the mast and sails were removed, and the oars stowed snugly in the bottom with the fishing gear, so that the next dory could fit in, like a cup on a stack. All this took time and meanwhile the wind had risen steadily. The best dorymen came last, at the time of greatest danger, and the captains always waited for them anxiously. Francisco Battista, the First Fisher; João de Oliveira, the Second Fisher; Francisco Martins, the First Fisher of the Azores and his compatriots Raul Pereira and Manuel dos Santos Rafael; and a dozen more including old Manuel de Sousa from Fuzeta and his son, and Antonio Rodrigues, from Fuzeta—veteran of forty-two campaigns—for these he waited, day after day, and watched them as they came, always with their dories full, and the First Fisher's fullest of all. Sometimes I could not see how that man dared imagine he could keep his dory afloat, though he took every possible precaution. He would sail back for miles, one split and cold-furrowed hand grasping his steering oar and the other the mainsheet, his lips split, his hands like red ploughed fields, torn with the constant handling of his lines, raw with the cold, sodden with sea water, his fierce and fearless eyes always on the sea, his immense skill sailing the dory which under any other man would have sunk long ago.

"He has never sunk a dory," Captain Adolfo said,

THE PROBLEM OF BAIT

watching him carefully lest the record be broken. "Not yet!"

This was an unusual record. Most of the good fishermen had had to swim at least once, when over-filled dories sank beneath them or overturned in the sea. Captain Adolfo never knew when the First Fisher's record might not be lost, and his life with it. Yet day after day he came back, the sea lapping at the gunwales and every breaking wave within millimetres of coming over his stern. Sometimes he gave his lines and some of his fish to relatives who fished near him, or to the green fisher Manuel Lopes da Silva, who was his protégé and whom he always brought back in the thicker fogs. When he knew a full dory would be in more danger than usual, he had a nephew, or cousin, or young da Silva in close company—not to save him; he did not think of that—but to save his fish. Frequently he would stop, bail furiously, gauge the danger of the rising sea, and discard more fish to the friends in company, or carry on. Back alongside, I often saw that Captain Adolfo was more worried about the First Fisher's safety than that remarkable man was himself. He would go on gaffing up his fish, his dory at last alongside and still in the gravest danger, working with savage speed and indomitable energy while the broken water swept into his dory. As long as he could get more weight of fish out than weight of sea washed in, he was safe. Yet often it was a near thing. Other dorymen, back earlier, had made quite a business of gaffing up their fewer fish. Not he! He just got on with it, all his energy directed to the job in hand, his fierce eyes upon the sea as if he could quiet it by his own inflexible determination. Day after day he came, his relatives bringing his lines and gear, and in his dory, only his fish and himself; sailing and bailing, pitching out gallons of water and cods' blood as his overladen dory wallowed in the sea, staggering on. Day after day he came, he and the other good dorymen, jumping in the tumult, thrown up on the crests of seas all round the bows for they always worked to wind'ard, coming to the ship well ahead of her so that they were not drifted past and into greater danger; each waiting his turn and his chance to get alongside. Often it rained, a hard, merciless rain, as the sea rose, and the little red dories leaping in the spray looked like overladen ants toiling in a surf. Yet one by one, skilfully handled, they slipped in under the bows, while the ship rolled heavily and the seas swept across her decks and the wind howled in the rigging. The *Argus* tugged at her anchor, plunging and jumping like a big dory herself, pitching violently until her forefoot was right out of the water; but slowly and steadily the nests of the recovered dories mounted, while the dorymen and the deckboys, the mates and the engineers and the cooks, toiled at the tackles.

"Hoist away bow! Hoist away stern!"

"Easy all!"

No other shouts. One by one, the sodden dorymen, watching their chance, leapt to the rail and climbed inboard, as their dories emptied and were hooked for lifting. Up they came, as she rolled, with speed but with infinite care. I saw the First Fisher come over the rail after bringing back his thirtieth full dory—he often filled twice in a day, or even in half a day—bailing for his life; his dory even when empty still in great danger, for the sea was so high that sometimes the frail red box hung for a perilous moment above the schooner's rolled-down side. I saw the First Fisher come over the rail and, with a foot on the swifter, give a last fierce look at the tempestuous and deceitful sea which for so long had been his enemy and would be again on the morrow.

For a second, he hung there in silence, looking; but his fierce eyes seemed to be saying, "Defeated again, you implacable wet hell! Defeated again!"

Yet for how long?

And then the waiting in the fogs. This was always trying. The same men were always last. Until the last tiny triangle of the last dory sail loomed ghostlike in the fog and, after a hundred false hopes had been aroused by shapes which were not there, its reality was quite established; until the last doryman was alongside, Captain Adolfo knew no peace of mind. Peace of mind? I could not see how he could ever know that—he, or any of the captains. I wondered that they did not suffer from nervous breakdowns, for they could know no peace. The successful conclusion of one voyage—if it were successful—hastened only the preparations for the next. The recovery of all the dorymen assured their exposure to the same risks the next day, the next week, the next month, next year—and all the years.

A man in command of a codfishing schooner, it was clear, had to be a man with great strength of mind and force of character. There was more in this kind of seafaring than I ever had imagined. As our time upon the Grand Banks drew towards its end, I looked forward to the continuance of the campaign in Greenland waters. There, they told me, it really would be tough. Compared with Davis Straits, the Grand Banks were a playground.

CHAPTER EIGHT
INTO NORTH SYDNEY

> But paths so harsh and of such length of days,
> Alas, they could not easily retrace.

"WE HAVE bait for only one more day," Captain Adolfo said, staring across at the *Creoula* anchored near-by. At last the weather was good and the cod abundant. That last day of May we had taken a good 200 quintals of fine fish. Even Captain Adolfo admitted that it might be 180. It was night then, and moonlight. On the fore deck the cluster lights illumined the splitters and the throaters, the gaffers, trolley-pushers, liverers, removers of tongues, and all the others who made up the industrious crowd. The Little King was passing salt down below, out of the way. The First Fisher, the boatswain, old Jacinto Martins, old Manuel de Sousa and his son—these, and a score more, were flailing knives and splitting fish as if their lives depended on it. Near the bow, flames rose from the narrow funnel of the oil-fired boiler which supplied steam for the cod-liver oil presses, and the belching flames were reaching halfway to the bright anchor light suspended in the fore rigging.

The *Argus* lay so quietly that the chain no longer champed at her graceful bow. A cable or so distant on the starboard beam, the *Creoula* was a picture of beauty in the moonlit sea. She, too, lay quietly, and her oilskinned dorymen could be seen busily splitting fish on her fore deck, which was a counterpart of ours. Round the horizon were the working lights of seven or eight more Portuguese hand-liners, and a Canadian or two, all busy cleaning and stowing fish. Among them were our company's ships *Gazela* and *Hortense*. We looked after the *Hortense,* and had given her bait. Captain Almeida in the *Creoula* looked after the *Gazela,* whose youthful master had been his mate the year before. The *Argus* was senior ship in the little fleet, and Captain Adolfo a sort of commodore. His duties rarely worried him, though on occasion they were very real.

Now, saturnine in his black fur cap and dark clothes, with an admirably assumed expression of critical disapproval he looked moodily round the familiar scene. The *Creoula's* working lights went out long before ours. Captain Almeida had claimed only 150 quintals for the day.

"*Creoula* has no bait at all," Captain Adolfo said. "We will share what we have with him in the morning. Then we will fish and, tomorrow night, make towards North Sydney in Nova Scotia. Captain Labrincha with *Aviz* will come with us and, of course, Captain Silvio."

There had been other signs of the approaching end of our brief Banks campaign besides the disappearance of the bait supply. For some days, the deckboys had been picking out the smaller cod and, after lightly salting them, putting them to dry in the wind. A dory had been stowed on top of the charthouse to house these cod which were to be, I learned, food for the visit to port and for the trip to Greenland. The cook, with his assistants, had been busy laying up huge stocks of cod steaks, both salted and fried. Since he could never know whether an extra fifty or more dorymen were going to be back for lunch or not, he always kept large stocks of fried cod steaks in a small pantry off the galley. Now these stocks were larger than ever. We had also put a few halibut in the refrigerated room. We still had fresh potatoes, shipped in Portugal but originally grown in Maine, and we still had some fresh fruit from the Azores and St. John's.

Had we then been in need of fresh food, we could have had some from the assistance-ship *Gil Eanes* which was on the Banks. She was hurrying among the ships, picking up men who were seriously ill, of whom there were fortunately very few, and delivering mail, parcels from home, fresh potatoes, lubricating oil, and such things to the ships who needed them. Many of the ships had not been into any port since they had sailed from Portugal. Besides those without refrigeration (for which a visit to port would have no particular advantage) many of the smaller ships had their supplies of bait brought them by larger ships belonging to the same owners. Only about half the fleet had gone into St. John's. The other half especially appreciated the visit of the *Gil Eanes:* many had already been two months at sea.

The arrival of the *Gil Eanes*, with Commander Tavares de Almeida as the representative of the Gremio and a sort of Portuguese admiral of the Grand Banks and Greenland, was an important event. We knew she was on her way because we heard messages being sent to members of her crew, during the broadcast Horas de Saudade. Then one morning she suddenly cleared the air of the codhunters' chatter, announced that she was on the Banks, and requested all ships there to communicate their needs by radio telephone. It was most interesting to hear all the captains speaking, reporting —very guardedly—how they had been faring. It took several hours to hear the reports from all the ships, trawlers as well as hand-liners. The fleet surgeon in Captain Silvio's motorship *Elisabeth* reported on his cases, and so did the trawlers' surgeon in the *Aguas Santas*. The fleet surgeon for the hand-liners had been in the *Elisabeth* since Lisbon, with a male nurse, and there were other male nurses distributed among the ships. It was intended to have one in each ship with a crew of fifty or more, but there were not enough to go round. Captain Adolfo looked after his men by making use of the male nurse in the *Creoula* and, when it was necessary, consulting the surgeon in the *Elisabeth* over the air. Then, on the surgeon's advice, either our second mate gave any necessary treatment, or we went over to the *Elisabeth* and sent the sick aboard by dory.

Medical work of this sort was done without regard to nationality. The *Gil Eanes*, indeed, had been delayed on her way across the North Atlantic by an urgent call from a French cargo-liner with a man suddenly taken ill. The surgeon diagnosed acute apendicitis; the *Gil Eanes* and the cargo-ship hurried towards each other; the sick man was transferred, operated on, and later, landed at St. John's. The ancient *Gil Eanes*, which was nothing but a little 2000-tonner built as a short-haul tramp, was the only hospital-ship on the Banks or, indeed, in the North Atlantic, and she was a most useful vessel.

We had not the good fortune to see her while we were on the Banks. No matter, we should see her in Greenland. Meanwhile the cheerful voice of Commander Tavares de Almeida was a welcome change on our radio, and it was useful to be able to communicate through the old ship with the rest of the world. Normally, we had only the telephone, and did not address shore stations. The *Gil Eanes* was in touch with the whole world, constantly, and carried an expensive and comprehensive radio installation.

Whether the *Gil Eanes* brought us a change of luck or whether the cod, out of their natural perversity, chose to assemble in strength just as the bigger ships were out of bait and hurrying away to replenish at North Sydney or St. John's, nobody could say. Certainly our last two days were the best two we had known, and they added at least 400 quintals to our small cargo.

"Wait and see; our Captain Adolfo will get a full cargo yet," said César, the engineer, as we watched the First Fisher coming back that last day with a second full dory, and Francisco Martins circling the stern in dory 51, still under sail, and also loaded to the gunwales with fat cod.

Francisco Martins somehow always seemed picturesque. He was a fine figure, six feet and more in height and splendidly built. He had a fine, open face, bronzed by exposure, and his features were regular and strong. He had a nice sense of colour which was plainly evidenced in his outer garments and in his dory, and the little red boat, coming home in the evening with her blue sails full of the Atlantic breeze and a curve of white water at her bow, always looked graceful even when the sky was black all round her and the sea leaping. Doryman Martins did not deliberately splash colour upon his dory or himself. The sails were blue to distinguish them, and his striped trousers, check shirts and tartans were no more brilliant than twenty other dorymen's. But he wore them with an air, and he was one of the most cheerful of the dorymen. Footballer Raul, who was also from the Azores, who earned his winter living by playing professional football extremely well, was another cheery soul. He often came back singing. He sang only in his dory, and not aboard the *Argus*. For some reason which I never discovered, it was not the thing to sing loudly aboard. I believe it was thought to bring bad luck, like putting the dory anchors in stern-first. This was a heinous offence and most unlucky, and so was spilling out any dory gear at the launchings and nestings.

On that last day, João de Oliveira had so many fish in his dory that he had cleaned them to prevent their weight from sinking the dory: in this way he was able to bring back two loads at once. He had gone a long way. The wind dropped, and he had to row. He was on fish, and therefore he did not care to come back with the first load. By cleaning the fish and discarding their heads he could stow more than twice the number. Salvador and the Star, worthy brothers from the Algarve and a cheerful, competent pair of first-class dorymen, came back towing half their long-lines behind them, still with the fish on the hooks, because they did not dare to take any more fish into their boats. Even the

Little King had three parts of a dory-load and looked pleased with himself. The *Creoula* also did well.

That night the dories were scrubbed out as they were emptied, and shifted aft. The faces were cut from all the big cod and salted in six large oil-drums cleaned out and cemented for the purpose. These would serve for food in port. Captain Adolfo, who abominated cod in general, liked eating cods' faces, and so did the mates.

As soon as the last dory was nested down, we got under way slowly to keep rendezvous with the *Elisabeth* and the *Aviz*. *Creoula* was already with us. Then, just as we got among the trawlers fog came down. We were over 400 miles from the coast of Nova Scotia and had to cross the St. Pierre Bank and Banquereau to get there. We had not fished on these Banks, and none of the hand-liners had done so. We had not sailed fifty miles before we were on a part of the Banks where the bottom permitted trawling, and the night was made dangerous by groups of large trawlers crossing and re-crossing our tracks. Fog did not deter them from working any more than it stopped hand-liners, but it made passage-making across the Banks a very real nightmare. The wind was fair from the east, and we made good progress. We kept in touch constantly, by radio, with the three ships in company, exchanging information. Each had all the men on watch on lookout, except the man at the wheel. We had four lookouts right in the bows, four more on the for'ard house, and the others posted along either side. Captain Adolfo left the deck only to use the microphone, and that rarely. Once or twice the hoarse bellows of a great steamer reminded us of an added danger: once we heard the wash of a

INTO NORTH SYDNEY

mighty propeller and the engine noise of a motor-ship which sounded as if it came from a 20,000-tonner.

Sometimes we saw the white water breaking at trawlers' bows, and that was the first we saw of them, for they were all painted black or grey. They were usually in groups according to nationality, here a dozen Portuguese, there six Spaniards or ten Frenchmen. They were not so great a danger so far as collision was concerned because they were dragging very slowly, but we did not wish to take undue risks with them, or to get mixed up with their gear. They were all keeping watch on the same radio telephone wave-length as ourselves, and we relied on the comradeship between all the fishing vessels to extricate us from any sudden emergencies. We did not slow down: we were making only about six knots.

But the big steamers were a real menace. We had to cross the approaches to the St. Lawrence and the steamship tracks along the American coast from St. John's towards the south, as well as the route from Europe to Halifax and Boston. Once when the bellowing of some huge motor-ship seemed to indicate that she was coming directly towards us at the rate of many knots, Captain Adolfo looked rather anxious. But no note of anxiety ever came into his voice, or the voice of any of the captains who were with us. They were not afraid of fog. They would have died of heart-failure long ago, if they had been. Captain Adolfo told me of nightmarish journeys in wartime, on this same run, when once four ships in company had been entangled with a great convoy coming out of Sydney bound for Europe. There must have been a hundred steamers going in all directions. The *Argus* twisted here and there, dodging in and out of the unseen columns, now under the bows of a big oil-tanker, now narrowly avoiding an American liberty-ship, or setting her big stays'ls frantically to get speed enough to avoid collision with some blundering tank-landing ship, come from Pittsburgh, and in the hands of a mariner who had acquired his title after a six weeks' course somewhere on the Great Lakes. Captain Adolfo shrugged his shoulders and grinned at the thought of it. The idea of dodging convoys in heavy fog seemed to delight him.

Once, he said, a big liner came so close that his schooner—he was in an old three-master named the *Neptuno*—was actually caught up and pulled along beside a 30,000-tonner by suction alone. Some of her dories were smashed, and he could hear the men high above him talking excitedly about his fish. The liner had gone on, leaving the *Neptuno* rolling violently in her wake. It was only by the grace of God that the liner had swerved in time and passed alongside the schooner, instead of through her. He had read the name. It was *Berlin*. Many Bankers had been knocked down and all hands drowned in such incidents as these.

The fishing schooners did not always come off second best. Sometimes in fogs they themselves sank other vessels. The *Infante de Sagres*, a shapely three-master from Oporto, had sunk two other ships and was still on the Banks, as good as ever. The danger seemed more threatening than it really was. The Portuguese captains were as good navigators in fog as any seamen could be. Some of the trawlers had the additional safeguard of radar and sent reports from time to time, when they were close to us. One was commanded by a friend of our captain. This trawler was without radar but Captain

Adolfo found her, hove-to close by, and sent the First Fisher over in his dory with a present of some Azorean pineapples. I thought this a good act. Captain Adolfo was always doing little things like that.

By night the fog cleared, showing our consorts in the moonlight, *Creoula* and *Aviz* looking most handsome under all sail. The *Creoula* was on big Gloucester lines like ourselves, but the *Aviz*, built by the Monica and of modern design, had a distinct trace of the caravel. She was a beautifully built four-master with a short modern rig and no bowsprit. She rode the sea like a graceful white lady, and even with her short rig, kept up with us well. Captain Silvio had set what sail he could on the stumpy masts of his motor-ship, and staggered along. The wind blew up fresh from the south-south-west in the night but the sea was flat, though we were not then in the shelter of the land. I wondered if this could be the mysterious place where Diogo de Teive and his son João had been, in 1451. They, too, had known strong westerly winds, with a flat sea. We were past the Banks but could see no land.

The fog came down again as the second afternoon wore on and the evening came. It was the first week in June. This, I believe, is a bad week for fog in that area. So, apparently, are the other fifty-one. The night was dirty, with heavy rain and wind, but the sea was still quiet and our progress excellent. We were well in the traffic lanes by then, sailing on blind, the barometer down, the wind rising. Ahead of us was the rockbound coast of Cape Breton Island. After midnight it cleared and, in the middle watch, we made the land. We were anchored in the pretty bay off North Sydney before noon, with our consorts close, and the schooners *Con-*

destavel and *Santa Maria Manuela, Argus's* other Portuguese-built sister with them. Three hand-lining motorships were alongside at Leonard Brothers' wharf, taking bait, and several others were waiting. The sun shone, and the rigging was promptly festooned with long lines of the dorymen's oilskins, freshly oiled for Greenland and put to air: the top dory on each nest sported a drying sail. Small buoys sewn from canvas which was later painted in bright colours hung everywhere to dry (these were wanted for the Greenland campaign), and the dorymen were busy removing the hooks from their hundreds of snoods, testing the hooks and examining them, and hanging the long-lines to dry.

At anchor near us was the new motor-ship *Soto-Maior*, a seventy-doryman, on her maiden voyage, a graceful and able vessel. Her master had been notorious for his pessimism on the Banks and was sometimes a harsh critic of his dorymen. Now I paid him a visit and found him a pleasant fellow, all smiles, and proud of his fine ship and crew. The *Soto-Maior* hailed from Figueira da Foz, a port which had been sending good vessels to the Banks for at least four and a half centuries. Her captain showed me the electric dough-mixing machine in the well-equipped galley, but complained that it made the bread too well and everyone ate too much of it, including himself. The quarters of the new motor-ship were pleasantly decorated. She was modern and well-equipped in every way, though her stacks of dories and her horde of check-shirted, dark-jowled dorymen were in accordance with tradition. The *Soto-Maior* had a crew of ninety-three.

The motor-ship *Vaz*, which was also in the port, had a crew almost as large. She had as mate an ancient codhunter who had long been master in his own right, and then retired, only to sicken of what seemed to him the aimless life ashore at Ilhavo, and seek a berth afloat again. He did not care for the responsibility of being master, though he had several sons who were masters in the fleet. It may be (as was hinted more than once, though never by Ilhavans) that Ilhavo was something of a matriarchy. Certainly the women there had been left to their own devices while their menfolk were at sea for six to nine months of the year, for at least four and a half centuries. The seafaring traditions of Ilhavo, indeed, went back a good deal further than that, for it had always been noted for its sailors and its fishermen. There were critics who declared that the women of Ilhavo were so accustomed to the undisputed command of their homes and everything in sight, that they now ran Ilhavo and the menfolk too, and the only retreat a man knew was a ship. Whether this was so or not I don't know, but there were fishing communities in which such conditions appeared to prevail. At Nazaré, for instance, it was said the women took the fish from the men as they landed, sold them, and only doled out to the fisherman a few niggardly centavos for tobacco. Perhaps that old mate merely found the life ashore unbearably lonely when all his friends had gone to the Banks without him. The only matter on which I could find complete unanimity concerning the women of Ilhavo was that they were extraordinarily beautiful.

Every day one of the four captains who had arrived in company entertained his colleagues aboard his own ship. They took turns in this, but they generally had one meal each day aboard the *Argus*. These occasions were always interesting, from the moment the captains came over the rail. They took a great delight in their endless arguments and sometimes noisy discussions, which were always conducted in the most friendly fashion and never settled anything. Who had taken most fish? Who had the best dorymen? (or the worst). They had only to lift a hatch-cover and look, aboard any ship, to see how she had fared. But they did not do this. It was not playing the game. Their arguments would begin on deck, much to the delight of any dorymen who chanced to be there. These listened with the keenest appreciation.

None of the captains believed that the *Argus* could possibly have taken less than two thousand quintals, as Captain Adolfo declared. The *Elisabeth* had more than two thousand, but the *Aviz* had not done so well. The *Creoula* had about the same quantity as the *Argus,* whatever that was. The four captains often spoke of the history of Banks fishing, as they knew it, of the origin of dories, and of the so-called Gloucester schooner type (which they said was from the Algarve, and that may be true). As for dories, some thought the type had probably spread through the Mediterranean with the Arabs, whose sewn fishing boats on remoter stretches of the Hadhramaut coast, though much larger, still have a discernible trace of the dory in them. Captain Silvio thought the dory had developed from a beach fishing type. Indeed, many of the notable beach fishing craft used at Caparica, Costa Nova, Furadouro, Mira, Vieira de Leiria and elsewhere on the coast of Portugal showed some affinity to dories. Fishing craft very similar to dories had been in use in the Olhao district for centuries. Perhaps the word itself was derived from the Portuguese word for fishermen, the boat of the "pescadores."

INTO NORTH SYDNEY

Whatever the truth of the matter may be, these were interesting topics for discussion in the saloon of the schooner *Argus* where they were expounded nightly, with much emphasis of statement and of gesture. There was a captain from a little salt-carrier in the Oporto trade whose gestures became so vehement, when he was carried away by the strength of his convictions, that it was slightly hazardous to sit too near him.

As for our dorymen, they were given $5 or $10 each, as they wished, and, each day of our brief sojourn in port, numbers of them were to be seen wandering through the North Sydney shops buying things for their wives and children. Few bought anything for themselves, though the Little King bought himself a warm jacket, secondhand, for the Greenland fishery. The excellent seamen's mission, run at North Sydney by the British Sailors' Society, made the men most welcome, and was of considerable assistance to them especially with their mail. The men received letters at North Sydney and there was much pleasure when these were distributed. I saw old Manuel de Sousa delighted with a snapshot of two minute grand-daughters, dressed for a religious festival in Fuzeta. The rancho and the mission became, for the moment, full of letter-writers. The mission provided also plenty of hot water, and this was greatly appreciated.

Many of the fishermen paid interested visits to the small Nova Scotian line-fishermen in port, some of which were so small that they fished only with one dory. A Nova Scotian or a Newfoundlander would rig a bathtub as a schooner if he were asked to sail it. Some of the schooner-rigged vessels using Sydney could not have been thirty feet long. Our dorymen were more interested in their gear, and the manner of using it.

Then one day Leonard Brothers announced that fresh mackerel was available in quantity, and it was our turn to go alongside the little wharf and embark our bait. Most of the other ships had gone by then, having taken the previous season's mackerel. It was almost the middle of June, and our four ships were the last. But with fresh mackerel bait the captains considered they would have an advantage.

CHAPTER NINE

IN THE ICE

> By sea, how many Storms, how many Harms,
> Death in how many sev'ral fashions dres't!

THERE are, in summer, two possible routes from North Sydney in Nova Scotia to the fishing grounds in Davis Straits. Ships may go direct, by way of the passage round the western coast of Newfoundland and through the Straits of Belle Isle, or they may sail out clear of the Gulf of St. Lawrence and make towards the north, passing Newfoundland on its other coast. This second way is safer because it is usually more clear of ice. It is also considerably longer. Though the other ships taking bait from North Sydney had all sailed that way, Captain Adolfo was keen to use the shorter route if possible. Belle Isle is close to the southeastern point of Labrador. Even in June, the weather there can be very bad, with fog and gales. There is always some ice about and a ship reaching the North Atlantic that way has the further disadvantage, if she is bound for Greenland, that she must cross the whole of the Labrador Current. This current flows southwards down the western side of Davis Straits, and it is one of the chief means by which dangerous ice is brought to the North Atlantic shipping lanes. The ice drifts from the Labrador coast and from Baffin-land and the supply is unlimited.

But we were the last ships to bait and sail for Greenland. The captains were in a hurry. The morning we left, Captain Adolfo received news that the Straits of Belle Isle were clear of ice except for grounded 'bergs and some growlers. By that time, June 12th, many of the Portuguese schooners were already fishing off Greenland, and we had no time to waste. So we headed towards the west of Newfoundland with the wind in the southwest, fresh. It was a bright, clear day, and the little *Elisabeth* had difficulty keeping up. Indeed, we had to douse some of our big stays'ls, for she was dropping astern. All four ships were heavy laden, with freshwater and fuel tanks replenished at North Sydney as well as the bait supplies. Captain Adolfo was again commodore of the little party, and we bowled along splendidly in the comparatively smooth water in a loose formation, sometimes with the *Creoula* in the lead, and sometimes ourselves. When the wind dropped, *Elisabeth* went to the front; but generally we had a good sailing breeze. The weather was pleasant, for the moment, though there was still plenty of snow on the Newfoundland hills, and there was a fresh nip in the air.

Meanwhile the fishermen, taking advantage of the good weather, put the finishing touches to their new long-lines and the cork buoys they had made to mark them. It had not been the custom to buoy the shorter lines on the banks off Newfoundland as this was considered to be unnecessary, but often twenty or more dory-carrying ships were together on the banks in Davis Straits, and it was desirable that the lines should be marked. The banks there were much smaller, causing the ships to crowd together, and there might be anything up to eleven or twelve hundred dorymen fishing the same area from ships all in sight of one another. The long-lines could become snarled together if buoys did not indicate where they were.

But why, I asked, go to Greenland at all. The place was notoriously difficult for fishermen of any kind. Strong winds funnelled up and down Davis Straits, often rising suddenly and bringing up a nasty sea on the shallow fishing banks. There were only four or five banks worth fishing—Fyllas Bank, about forty-five miles by an average seventeen, off Godthaab fjord; Lille Hellefisk, off Sukkertoppen, about two hundred and fifty square miles in area; a strip of coastal bank to the north of that, some of it in territorial waters; and Store Hellefiske Bank, with a least depth of eleven fathoms, ranging from off Holsteinsborg Bay just north of the Arctic Circle, some hundred miles to the nor'rard, as far as Disko Bay, and covering about 3,000 square miles. This was the richest bank, I was told, because it was used as a feeding ground by enormous schools of cod which spawned in the fjords on that part of Greenland's coast. But much of the best of it lay inside territorial waters where the fishing was restricted to Danes and Eskimos. About half the rest was used by trawlers. Fyllas Bank was regarded as particularly dangerous because of ice, and both this bank and the smaller bank to the south, known as Dana's, were much fished by the Faroese, Danes, Icelanders and Norwegians.

THE QUEST OF THE SCHOONER ARGUS

Strong tides raced over all these Davis Straits banks. Much of the bottom was uneven, and this could cause a sea dangerous to dories. There was almost as much fog as off Newfoundland and, because of the nearness of the magnetic pole, small compasses might be most unreliable. To be lost in fog in comparatively still waters was bad enough, but to be adrift in a four-knot current might be fatal. Our experience during the last few days on the Grand Banks seemed to show that there were plenty of good fish there. If we could fill the schooner two and a half thousand miles from Portugal—quite far enough—it seemed odd that we should sail another one and a half thousand, and all of that towards the wretched north where our prospects were most dubious.

Captain Adolfo explained that it was unlikely that the whole fleet, or any of the larger ships, could fish full on the Newfoundland Banks unless the season was exceptional. Even before the big trawlers came there, when the Portuguese and French fleets consisted entirely of smaller ships with an average capacity of two hundred to three hundred and fifty tons, there had been many bad seasons when they could not fill. With the advent of large-scale trawling towards the end of the 1920's, the line-fishermen had to go further afield. Some seasons, perhaps, they could all fish their fill on the Grand Banks, using their lines in areas where the trawlers could not go. But it was impossible to count on this. Providentially, just when fishing prospects were becoming most doubtful on the Grand Banks, the Davis Strait fisheries had opened again. Because better hydrographical conditions, following what scientists call a "warm cycle," had brought the cod back in large numbers to

IN THE ICE

the Greenland banks, and much of the bottom there was unfit for trawling, the larger line-fishermen had been accustomed to take the greater part of their cargoes from Davis Straits for the past twenty years. They might fish full on the Grand Banks, but it was far more likely that they would not. The cod were usually more concentrated in Greenland waters. Hence the use of longer lines. If a ship had the luck to get on fish there, she could probably take vast quantities and, if the weather allowed, fish her fill in two months or so.

Captain Adolfo had done so on several voyages, taking as much as 14,000 quintals. He had never done so well on the Grand Banks. In good seasons, the *Argus* could fish an average two hundred quintals a day on Store Hellefiske Bank. Not for years had a ship taken anything like that quantity with hand gear, on the banks off Newfoundland. The Greenland fisheries made it possible for the larger line-fishermen to go on fishing.

Captain Adolfo knew the risks and so did the dorymen. They had to be accepted. For the sake of the cod the dorymen would go anywhere, and so would the captains. Greenland voyages had not added appreciably, if at all, to the loss of life or the loss of ships. Perhaps two ships a year were lost on an average from all causes. The majority were lost on the Grand Banks or on passage, not off Greenland. The loss of life had decreased sharply, but that was due almost entirely to radio. With radio in all ships, the dorymen, even in a fog on the Greenland banks, were much safer than the earlier dorymen had been, without radio, fishing on the banks off Newfoundland. A lost doryman in Davis Straits had a chance of reaching the coast in safety and landing there, for he was never far out. Many had been saved that way. Few had ever succeeded in sailing their dories from the Grand Banks so far as Newfoundland or the coast of Nova Scotia, though some had done so.

Ice was usually worst on passage, not on the banks which, curiously, became more ice-free the farther north the ships sailed. North of Fyllas, Captain Adolfo had sometimes seen no ice for weeks, apart from grounded bergs and a few odd patriarchs drifting in the current. He hoped the warm cycle would continue at least so long as he remained at sea: if it came to an end he would never be able to fill the *Argus*. The ice-free area up Davis Straits and in Baffins Bay was apparently still extending, and it was said that cod were to be found in Melville Bay. He had not been there and did not propose to go if he could help it. Store Hellefiske Bank was far enough, and there were still plenty of cod there. Since they had begun to fish the Greenland grounds, few of the Portuguese had failed to secure a fair cargo, except in seasons of prolonged bad weather.

Meanwhile our good progress continued. The weather seemed determined to remain good, and the Straits of Belle Isle were like a swift-running river. It was a river with more icebergs than we cared to see, though most were aground or appeared to be. There were also far too many growlers. These growlers—small icebergs—can be the most dangerous form of ice in the open sea, especially when the wind brings up whitecaps to hide them. Then it is often impossible to see them before the ship is practically on top of them.

The four ships in company kept good lookout. Sometimes we saw the *Creoula,* the *Aviz,* or the *Elisabeth* take a sudden wild lunge away from a murderous piece of ice right in her path. We often did the same thing.

But we continued to run splendidly with the fine fair wind, and the sun, going down blood-red over Labrador, promised fair weather on the morrow. The night was fine and clear, and the navigation lights shone brilliantly.

"Too good, too good!" said César, shaking his head. "We shall pay for this!"

As soon as we had cleared the straits, small icebergs, growlers, and pieces of old pack became much more numerous. That did not matter so long as the visibility was good. The nights were becoming much shorter, which was a help. Each ship kept the whole watch on lookout except for the helmsman and gave radio warning to the others of any particularly dangerous ice which might be sighted. If the ice threatened to become too bad, we hove-to until the morning.

On the fifteenth of June, heavy fog came down, and we had not gone very far that day before we were caught in a really dangerous field of ice. The sea, fortunately, was flat. The air seemed dead; and the day was savagely cold. The sails were run down and caught with a loose turn of the gaskets, to be handy for setting when it blew again, and the little fleet went on slowly, using motors. We had just crossed the latitude of 54 degrees North and were not far from the coast of Labrador. We were by no means clear of the Labrador Current yet. At first, there appeared no reason why we could not continue slowly, picking a way carefully past the more dangerous ice, but it soon became obvious that this was a field we were not going to get through, at least while the visibility was so bad. The only thing to do was to get out of it if we could, heave-to, and wait for the fog to lift.

THE QUEST OF THE SCHOONER ARGUS

But getting out of the field was not easy. Most of the ice was small and well weathered by weeks of motion in the sea, though some of the bergs were half as large as the *Argus*, and could easily have sunk her. The sea was littered with the horrible misshapen pieces of ice, many of them squirming uneasily, like big jellyfish. Many showed streaks of lovely pale blues and greens even on that grey and thoroughly foggy day, but we were not in a mood to admire their beauty. A brush with any of them might mean a stove-in plate or, for the wooden *Aviz* and *Elisabeth*, a crushed hull. All hands lined the rail—dorymen, deckboys, and the rest—each with any long pole he could lay his hands on, pushing away the floes and the small growlers as if his life depended on getting them as far from the ship as possible. Dory-masts, long oars, boat-hooks, fish gaffs, baulks of timber were pressed into service. The First Fisher and old Antonio Rodrigues were sweating like pigs, pushing away a nasty white mess which looked like a gigantic frozen mushroom and persisted in rolling over and over as they pushed it away. It always seemed to be coming nearer to the ship. Sometimes the scrunch of smaller ice could be heard along the sides. The water had gone greasy, like cold pea soup. Bobby, the little yellow dog, jumped up on the rail to see what the excitement was about, caught sight of a big growler and immediately leapt down again. Once the *Aviz* loomed up, almost on top of us. All hands were busy aboard her, too, pushing ice away with poles. It was a good thing we had such large crews. Somewhere in the fog not far away we could hear the sirens of the *Creoula* and *Elisabeth* wailing like banshees in distress.

César de Medeiros

The Trolley-pusher

The Excellent Chief Cook with His Pet Hen

Deckboy with a Big Cod

The *Argus* Lay Quietly

The *Aviz*, Sailing for North Sydney

Bound for Greenland

Meanwhile, Captain Adolfo, now lending a hand at the ice-shoving, now running to the telegraph or wheel, tried to extricate the ship from the worst of the ice. He had to be particularly careful that she did not back into a berg, even a small one. The *Argus* was strengthened against ice only at the bows, to help her to break through light old pack, or brash ice. This way and that, we twisted and turned, slowly. Again the *Aviz* loomed up suddenly, almost on top of us. This time we cleared her bow by less than a fathom. The sirens of the other ships sounded very close. Once we heard the voice of Captain Silvio announcing that he thought he was out of the worst of it, but almost immediately came a shout that he was in ice as thick as ever.

We got out, after twisting about for hours, and came at last into a slightly clearer area. Here we brought up, standing idly in the sea and drifting with the ice, waiting for the fog to lift, the watch on deck still lining both rails ready with their poles to push threatening floes away, and the watch below two-deep at the bows.

At midnight we were still there. At a late meal, César Mauricio spoke of the day in 1944 when, in exactly similar circumstances and almost in the same place, the three-masted schooner *Maria Preciosa* was brushed by an iceberg, ripped open, and sunk. There were three ships with her and these rescued the crew. No wonder these ships like to sail northwards in small groups. No life was lost from *Maria Preciosa*, though the ship sank quickly.

Meanwhile we were still in some danger of losing the *Argus*, or having to take off the crew of one of the ships in company. But just before four o'clock in the morning the weather cleared. Some grey, cold light showed a way out, and we took it cautiously. We had first to retrace our steps some distance towards the south. When at last clearer water was all around us, Captain Adolfo headed directly eastwards to get out of the Labrador Current and its load of bergs. We sailed east for the next twenty-four hours, losing the advantage we had gained by sailing through the Straits of Belle Isle. The *Hortense,* sailing up from St. John's towards the Fyllas Bank, reported clear water in the middle of the Davis Straits, but she said she could see many large icebergs to the east of her. These were in the Polar Current which, sweeping south along the eastern coast of Greenland, swings round Cape Farewell and then moves to the north, taking a great horde of icebergs and field-ice with it.

We sailed steadily eastwards all day with the sun shining brilliantly as if anxious to make amends for his previous absence. There were some large bergs almost like Great Barrier ice in the Antarctic, and once we took several hours to skirt a large field. This field looked like a great bay of ice which had drifted out bodily when the summer loosened its links with the shore. We could hear the ice grumbling and grinding as the sea, even on that quiet and sunny day, broke all along its weather edges, sending up clouds of spray which flashed and sparkled briefly before they fell upon the ice, to become ice in their turn. I watched all day with a vague hope that we might be granted some unusual sight, as was the brig *Renovation,* of Shields (E. Coward, Master), one bright summer's day almost a century before. The brig *Renovation,* falling in with a great field of icebergs not far from our path, saw upon the

IN THE ICE

field ice attached to the largest berg "two three-masted ships, having their masts struck and their yards down and all made snug," as the master later reported.

"To all appearances," his report stated, "they had passed the winter together on the ice. I took the spying-glass, and carefully examined them to see if there was anyone aboard, but could see no one . . ." These ships very probably were Sir John Franklin's lost *Erebus* and *Terror*, which had taken him on his ill-fated expedition in quest of the North-west Passage. These "having their masts struck and their yards down, and all made snug" certainly had broken away in the ice not long before and had not again been seen. Captain Coward later described them as ships of about 500 tons and 350 tons burthen, and added that the larger one was on her beam ends. It is a pity that he did not take a closer look.

We saw nothing. Nor did we pick up any parties adrift on the Arctic ice, as some ships had done. There was, for instance, the crew of the U. S. vessel *Polaris*, nineteen of whom became separated from their ship in 77 degrees North, and drifted on an ice-floe in the Labrador Current 1500 miles, to be rescued by a sealing vessel near the Straits of Belle Isle, nearly six months later. Sealing vessels themselves had lost many similar parties, and still may do so.

Perhaps the only thing we might see now, Captain Adolfo said, was an upturned dory. I asked him whether a wooden ship like the *Maria Preciosa*, which was almost full of salt when she foundered, might rise again when the salt washed out. Captain Adolfo thought not, though he had heard of wooden ships which, overwhelmed in the sea, had risen again when their cargoes had dissolved. The *Maria Preciosa,* however, had been pierced. He sincerely hoped that no such derelicts were in his path, or anywhere else in the frequented ocean.

Then it blew a gale from the east'ard: our motion was violent and the ice, though no longer thick, was now at its most dangerous, for small bergs awash in a big sea have sunk far more ships than large icebergs have ever done. We had to keep good way on in order to have the ship manoeuvrable, and were sailing nine knots. The *Creoula,* close under our lee quarter, still with one big stays'l set and straining at its sheet, was very beautiful with the sprays clouding her slim bows, and her sixty dories smothered every few moments beneath the sea and foam. She was riding the seas well, though she was plunging and rolling. The *Aviz,* looking more than ever like an ancient caravel with her big square running-sail set on the fore, was well astern. She, too, was making good weather of it, and excellent progress. Captain Silvio's motor-ship was not to be seen. No matter, he would come up when the wind eased. The lovely schooner *Hortense* was in sight, far down to leeward, bounding along like the thoroughbred she is. She is a first-rate sailer but, in the rising wind and sea, the larger ships soon overhauled her.

The loudspeaker had been telling us of the *Gazela,* already anchored four days on Fyllas and filling with cod, and of thirty other ships on that bank, and *Dana's,* also doing well. They must indeed be doing well if they talked about it. We drove on, now in heavy rain and once again in poor visibility. All round us lurked the little bergs, so hard to distinguish in the breaking water and out of sight altogether when they sank in

IN THE ICE

the troughs. Yet still we bounded on, all eyes straining. He who risks nothing, gains nothing, and a cod-hunter alarmed by ice and gales is better at home.

We sighted a stranger near the Arctic Circle, a little motor-ship looking incredibly small in that enormous sea. She was an Icelander, probably working from Faeringerhafen. We could read her name—*Hansobe*—and her number, which was R28H, by which I thought she was from Rejkjavik. An Icelander must be fairly well accustomed to wild motion at sea, but how anyone could live or work aboard that vessel in such conditions was beyond me. With some of her furious pitching she was flinging half her bottom out of the sea. Even our dorymen looked at her with mild astonishment.

We sounded, shortly afterward, and got forty fathoms. The sea was higher and more confused. There was no sign of land. We shortened down, heading east, the air full of the radio chatter of the other ships, *Vaz, Soto-Maior, Senhora da Saude, Rio Caima,* and the rest. No one was fishing. The sea was impossible for dories.

Early in the first watch the weather eased—it was still light—and the glass began to rise. We were shortened down to the lowers and the fore stays'l. Soundings then gave twenty-five fathoms. The sea was still tumultuous. Anchor in this? I considered that impossible, and thought we should keep under way.

"Haul down the sails. Leave the trys'l set," Captain Adolfo ordered: and when the way was off, we anchored. With her nose held by the cable, the ship leapt and rolled violently, but there was sea-room enough to drag all night and all week, if she were so minded.

THE QUEST OF THE SCHOONER ARGUS

Round us the little group of our consorts and the *Hortense* were also anchoring in silence. A first-voyage deckboy said something in a loud voice.

"Silence!" the second mate hissed. "Noise is unseemly when we come upon new ground!"

Near us were two large icebergs which we hoped were aground. The sea smashed on them, sending sprays high on their castellated sides. One was near enough for us to admire the lovely shades of green in its cable-long base. The *Santa Maria Manuela* was inshore of us, and we could see the masts of other schooners further off. Towards the land the sky was black, gloomy and forbidding.

CHAPTER TEN

THE GREENLAND CAMPAIGN—JUNE

> So strong a Current in those parts we meet
> As ev'n obstructs the passage of our Fleet.

THE trouble with the fishing on Fyllas Bank was that the fish were too small. There were also too many ships there. The weather was appalling and there was a great deal of ice about. The first night we were there, the deckboys jigged up about a quintal of cod, working with one jigger each, from the poop. These cod were all very small fish. Small cod are better than no cod at all, but the dorymen looked down their noses at fish less than thirty inches long. Smaller fish made more work, because they very much increased the splitting and cleaning difficulties. They did not seem to fill the ship as large, fat cod did. Their livers were small, more trouble to collect, and—because the fish were also rather thin—did not yield such good oil. Nor did these cod make such good food.

As for the ice, that same night we heard the *Santa*

THE QUEST OF THE SCHOONER ARGUS

Isabel warning the other ships of a large 'berg which was bearing down upon her. Later, this piece of ice—there was about an acre of it—came so close that Captain Simões had to cut his cable and shift anchorage, quickly. The *Santa Isabel* had the old-fashioned cables, with many lengths of stout cordage to help take the strain. Thinking the iceberg would clear him, the captain first paid out more cable; then, when the ice still bore down on his ship as if it had singled her out for destruction, his only course was to cut and run. So he lost the anchor and chain and all the rope cables, worth several hundred pounds, and his sailors were busy all that night getting up fresh cables and preparing the other anchor. There was a large iceberg very close, and we hoped it would not begin to drive down on us. It seemed, however, to be aground and to have no movement apart from a gentle rocking from one side to another, like an enormous white elephant rocking on its tether, asleep.

The next day we were not able to launch the dories, for the weather continued to be quite impossible. The sea on the bank ran whitecaps and deep furrows, irregularly. No one could fish. Our *Argus* rolled, bucked, jumped, lurched, and danced about as if she hated the place and wanted to part her cable, to drag anywhere out of that benighted area. Her motion was so violent that it was difficult even to sleep and to eat, and the dorymen surveying the cold murk and the small fish jigged up from the poop, spoke pessimistically of returning to Lisbon in October or November. A few said we should have done better to have stayed on the Grand Banks, declaring that the schooners which did so would be full and home long before we had a chance to fish full. But fishermen, like farmers, are pessimists everywhere. By midnight it was bright and clear, and full daylight, of course, though Fyllas is a little south of the Arctic Circle, and the coast of Greenland began to show. It was a black and mountainous mainland, pitted with the scourings of many glaciers and deeply indented. The grim skyline was serrated in wild contortions, as if the Creator had not been pleased when He put it there.

The first day we could fish, we succeeded in taking little more than a score quintals of small cod. At four in the morning it was bright, clear and sunny; and the dories were away at twenty past the hour. But by seven a bank in the south'ard foretold wind from there, and a wild black line was forming under the clouds massed along the horrid Greenland hills. That morning we could count twenty-seven ships fishing together on Fyllas, almost all of them Portuguese. When the weather looked threatening, Captain Adolfo, fearing the onset of the dreaded south wind, consulted his brother, and Captain Silvio, and a few other particular friends (all of them most experienced Greenland captains) as to whether they should hoist the recall or not. The opinion that we should do so was unanimous. Looking somewhat disgusted, Captain Adolfo emerged from the companionway and ordered the sugar-bag flag to be hoisted. Some of our better dorymen, who always sailed for miles, had scarcely had time to get their lines properly down. The hoisting of the recall, of course, did not mean that they had to come back immediately, but only that they could come when they were ready. But the dorymen could read the weather signs as well as any captain, and they did not want to be drowned. They all particularly dreaded any wind coming from the south, for they said that sometimes the winds from that direction came on suddenly, with little or no warning, and almost at once set up such a tumult on the water that a loaded dory could not survive. Many dories had been lost in that way.

Yet it was after eleven before the best dorymen were back; well after, before the Battista family came bounding across the breaking sea, and, not far away, João de Oliveira. Even these experts had not been able to fill a dory with their first long-lines, and they dared not stay out for a second haul. Wind and sea were steadily rising, though by the grace of the Lord, there was not then any sudden onslaught from either.

"We had the warning," said Laurencinha. "When there is warning, the wind does not come fast. No warning, watch out!"

He grinned as he rigged down his dory. Meanwhile Captain Adolfo, disgusted with the weather and the smallness of the fish, was getting under way again. If he could not fish, at least he could use the fresh south wind to sail towards the north: he had already had enough of Fyllas. The *Argus*, with the *Creoula* and *Elisabeth* in company, bounded north through the long afternoon.

"We will try the next bank," the captain announced. "It cannot be worse and it might be better. At any rate it will not be so bad for ice, and there will not be many ships there. Also there might be a better class of fish."

Far astern of us, we could already see some of the other ships following the example of the experienced trio. As for the ice, it seemed odd that we should go north towards the Pole to avoid icebergs, but it was explained that the bergs on Fyllas and the southern banks were, in the main, carried there from eastern Greenland, after having been swept round Cape Farewell and carried up in the north-setting current. This current dissipated itself as it flowed northward so that by the time a ship reached the Arctic Circle there was noticeably less ice, and by the time she reached Store Hellefiske Bank, there were almost no bergs at all. Those coming down from the north were swept into the Labrador Current, going south down Davis Straits, and those from the south did not get far past Sukkertoppen.

Meanwhile there were quite enough about, as well as a host of dangerous growlers, and navigation was still exceedingly dangerous. The visibility was good. We had good way, and plenty of lookouts, since there were so few fish to salt. We came to anchor again at nine in the evening, in nearly fifty fathoms of water, on the bank known as Lille Hellefisk. The schooners *Rio Caima* and *Antonio Ribau*, and the little motor-ship *Terra Nova*, which had followed us, anchored not far away, and we all proceeded to roll and jump violently in the high sea which was still running. The *Terra Nova* was a tiny thing of less than four hundred tons, hailing from Figueira da Foz, and recently converted from a three-masted schooner. With a steel superstructure built up on a wooden hull, her bowsprit sawn off and her good masts replaced by a pair of stumpy sticks, she looked odd, and I wondered why the rigging had been taken from her. According to the afterguard in the *Argus*, queer things were sometimes done to ships from Figueira da Foz.

Our first full day on Lille Hellefisk Bank was good, though it began with a threat of southerly, and Captain

THE QUEST OF THE SCHOONER ARGUS

Adolfo would not launch the dories until seven, when conditions were settled. By mid-morning the weather was bright and sunny, the ship lay almost at peace upon the flat water as she had never done on the Grand Banks, and the canary sang in the saloon skylight. But the grim coast of Greenland and the icebergs all round took away any illusion of summer, and the horizon was contorted by mirage and flung up in squares and parallelograms, as the Great Ice Barrier is in the Antarctic.

"A bad sign," said the mate. "Mirage is no good. We shall get wind again."

Meanwhile we had a dory back, laden with good fish, by ten o'clock, in spite of the late start; and the mate and the second mate jigged up a quintal of fat cod from the afterdeck in twenty minutes. It looked as if Captain Adolfo's move to the north had been a wise one. By mid-morning, indeed, the big *Milena* and two more motor-ships, including the *Capitão Ferreira*, had joined us. We watched them hoist out their dories as soon as they anchored, and soon the sea was alive with the little red and yellow boats skimming in all directions, some under sail, some oars. The majority went long distances from their ships before shooting their lines. Some of them were close to us (though none of our own dories was there) and I watched their method of procedure.

There was a considerable movement in the surface water across the bank, and I wondered how they could keep their thousand-hook lines clear. I saw that each doryman, as he came to his chosen spot, first threw out a grapnel to which a light buoy—generally flagged with his personal flag (a bit of painted canvas, or something of the kind)—was attached; then, as the current carried his dory along, he paid out his long-line from a tub in the bows, watching that each snood with its baited hook went down clear. This was assured, and could only be assured, by the manner in which he had first carefully coiled the line in the tub, so that it would run out clear and never snarl a single hook. He paid line out steadily and, when he came to the end, perhaps half-a-mile or so from his buoy, he secured the end of his long-line to the anchor of his dory, and dropped that. Then both dory and line were secured.

While his long-line was down, he fished by jig or baited hooks on hand-lines, and the number of fish he took in either way was a good indication to him how his long-line might be doing. If fish were plentiful, he would begin to haul in his big line when it had been down two and a half hours or so; if fish were not plentiful, he allowed it to stay down longer. He hauled in his long-line by first lifting his dory-anchor, and then, taking off the end of the long-line and securing it in the tub, he began to haul away, hand over hand. Since this was extremely laborious and often most difficult work —for the bottom was very rocky, the water deep, and the tides strong—it took a long time. Sometimes protruding rocks on the bottom tore away hooks or chafed through snoods; sometimes he would lose his grapnel and part of the long-line, and try as he might, could not recover it. As he brought the hooks to the surface, he would fling the hooked cod skilfully over his shoulder into the dory with a flick of his powerful wrist, not even looking at them, except as they came to the surface, when he could see whether they were hooked well enough to be hauled inboard without gaffing. Many were not and these he gaffed in, or put a hand down and hauled in bodily, if they were small. Meanwhile he was skilfully coiling the line in the tub again. The tub remained right in the bows, where the line was ready for its next use.

The motion of the dory even in the quiet sea was considerable, and the effect on a man's hands of hauling so constantly on hard wet lines in the Arctic sea may be imagined. All the men wore woolen gloves, and most of them hauled with rubber protectors on their palms. But to keep warm was quite impossible. Their hands were invariably split and covered with sea cuts—painful long openings in the hard skin which usually did not bleed but never healed, so long as the hands were wet daily. Hauling in the long-lines must have been purgatory.

When the long-line was all in and the grapnel lifted, it would immediately be laid again if the fishing proved reasonable, the doryman baiting those hooks from which the bait had gone, as the line paid out. If, in the doryman's opinion, his first "shot" had not brought in enough fish, he set off for another spot, perhaps miles away. There he would repeat the process as before, and he was a poor doryman who could not "shoot" his long-line three times in a day, if the weather permitted. Many men—the First Fisher among them—always tried to sail their lines out, partly across the tide, and never permitted the movement of the water to lay their hooks for them.

The First Fisher explained that he did this in order to keep his snoods clear and to have all his hooks well streamed out along the bottom. If, he said, he just paid out line with the tide, there was a far greater chance that the hooks would be twisted on to the long-line itself, and so snare none but cod determined to be caught. Clear lines meant better fishing. I gathered that his opinion of dorymen who merely let their dories drive with the movement of the surface water, was not high. However, some did that: there was only one First Fisher.

That day many lines were snarled and, when everything else was done—which was very late, for there were considerably more than two hundred quintals to clean and salt away below—the poor dorymen spent hours clearing their lines and preparing them for the morrow. I saw Captain Adolfo helping an old Azorean with his lines, and the young mates and the cabin boy were helping other men. It was after eleven when the fish-salting was done, and though the constant daylight made an appreciable difference to the men's ability to remain awake and energetically at work, it did not make turning out at four A.M. any easier. Few of the Fuzeta men had snarled their lines, but they found work enough tending them. The First Fisher always looked at every hook and carefully baited each one, especially with crustaceans when he could find any.

"I must give Mister Cod a good breakfast," he said. He did not seem to share the general contempt for codfish. Some men simply flung sliced herring pieces and bits of cod roe, and similar scraps, on to their hooks. Perhaps that was another reason why the First Fisher did so well. Everything he did was done with great thoroughness.

On Lille Hellefisk the lines brought up considerable numbers of a repulsive fish with a slimy spotted skin, cold murderous eyes, and a nasty mouth full of strong teeth. The dorymen called them "catfish." The scientific name is *Anarhichas Minor*. These they cut up for bait, as they did any halibut they caught, though some halibut found their way to the galley. We took no haddock, though these were plentiful on the Grand Bank. All dorymen had to get as much bait as they possibly could, irrespective of the ship's supply. It had been the same off Newfoundland. There was not bait enough to supply all the hooks all the time, and there never could be. The covers over the refrigerated chambers were chained and padlocked, and these were the only compartments which were ever kept locked while the *Argus* was at sea. The store-room for the wines and brandy was always open. If a doryman came back with a full dory, he was allowed a little additional bait, but it was always counted out most carefully.

We had then, in the beginning of the last week in June, four good fishing days one after the other. Each morning we made a slight shift of ground, sometimes only a few miles, sometimes eight or ten. Then the dories were launched, and they fanned out for miles. We always knew how good the day was going to be by the number of full dories which came back by midmorning. If a dozen were back and gone again by then, it was bound to be a fairly good day; but if Captain Adolfo saw many dories shifting ground after the first haul—generally they shifted about 9:30—he was most pessimistic.

"No cod! No cod!" was the cry, and the lament would be sent out over the radio for all the fleet to hear; and this despite the fact that the second haul might bring in prodigious quantities, and sometimes did. Sometimes almost all the dories would do poorly for their first two hauls, and then on the last of the day, take two hundred quintals. Why the cod congregated in a particular area no one could say. The spectacle of ships close together, with altogether different luck though they had dorymen of equal skill and energy, was commonplace. But there were masters whom good fortune seemed to favour specially, and Captain Silvio, in the *Elisabeth,* continued to be one of them. He was often in close company, but we did not often have his luck. When we had six or seven dories alongside, gaffing up their fish, Captain Adolfo would count ten through his binoculars alongside the *Elisabeth,* and remain disconsolate. Not that Captain Silvio ever admitted he was doing well. On days when his fishing was good, he used to play music on his loudspeaker all day long, at full pitch, and whenever he switched on his microphone we could hear it. Then we knew.

"Ah! Miséria!" was the general lamentation aboard the *Argus.* "That fellow on fish again! Why should the fish choose him?"

But the pessimism was not real, nor was the envy. On the whole, *Argus's* dorymen filled their dories more full before they came back, and everybody knew it: a dozen of our best fishers with their dories full must have brought as much fish as twenty who rowed back "full" to some other ships. Many dorymen, passing us to go back to their ships in the mornings with what they considered fair loads, still had nine inches of freeboard. Many captains, indeed, forbade the men to fish their dories over-full, for fear of foundering. An inexperienced doryman, finding himself on fish with some already in his boat, might throw too many into his dory over his shoulder as he hauled in his long-line, and not notice the water beginning to lap over his stern until he suddenly found himself in the sea. If he was not a very lucky man, that was the end of him. We had sufficient demonstrations of just how easily that could happen, even with the most experienced men. Some of the older men would not pay attention to safety. With them it was fish first and safety last, and nothing would change them. Not that they took foolish risks; but a high degree of danger they certainly accepted.

On one of these good fishing days, the morning calm and almost beautiful, the sun out so strongly that Captain Adolfo had washed the ship's two dogs, when a good many dories had already come back with very full loads, a puff of air came suddenly from the south and grew quickly, bringing a lop up on the sea. Many of the dorymen were rowing back. The ship at the time was tide-rode, lying across the wind. This put the Azoreans' side to wind'ard, and the sea began to slap and fret there, lightly at first but, very soon, to a dangerous degree. A good Azorean doryman had just put his dory alongside there—it was Mariano da Cunha, in dory 35—and he began to gaff up fish. But the lop of broken water gently lapped into his dory, and down she would have gone like a dead cod's backbone if the alert Captain Adolfo had not immediately grasped the bow grommet with a large boat-hook, and shouted for César—also always on the alert on such occasions—to do the same thing aft, which he immediately did. The deckboys were still down below working in the salt, making another compartment ready for fish. At this moment, several Algarvians were returning on the other side of the ship, and the mates and cooks were looking after them. Mates, cooks, and the assistant engineer ran across the deck, deckboys streamed up

THE GREENLAND CAMPAIGN—JUNE

from the hold, and all grasped boat-hooks and chain-hooks to help dory 35 to keep afloat. Meanwhile, the doryman after trying desperately to gaff up enough fish to prevent his dory from sinking, was floundering in the icy water. A fellow islander, very fortunately, was close alongside but out of the lop, and—also very fortunately—without his dory so heavily laden as Mariano da Cunha's had been. This doryman was able to manoeuvre alongside the sinking dory.

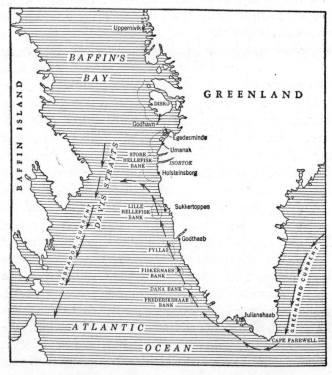

Greenland, Showing the Davis Straits Fishing Banks

First he rescued the tub of long-line, which was beginning to drift away in the sea. That safe, he rescued the doryman himself. This he did with difficulty, for the heavily-clad man was helpless after his immersion, and an appalling dead weight in the water. Several of the Algarvian dorymen, hearing the commotion, leapt inboard over the rail and ran across the decks to help. One of them jumped down into the dory, which was awash but not sinking any further, and began furiously to gaff up fish. Meanwhile the fish were washing out and the dory's gear was streaming back astern. As quickly as they could, those who were holding up the dory with their hooks passed it along until it was below the hoisting tackles. These were adjusted, the strain of the dory's water-logged weight taken upon them, and other dorymen began to bail furiously. The dory was saved. Dories which had already emptied rowed quickly back to retrieve the fish and gear. Doryman da Cunha was hoisted inboard, went below, changed his clothes, came on deck, and, asking whether I had photographed the incident, got on with the job of helping the other dories. He was cleaning fish until nearly midnight, and off in his dory as usual the next morning.

But if that slopping sea had come into his dory *away* from the ship, he would have stood little chance. Any full dory which tried to pick him up would have sunk herself, and on deceptively quiet days like that, every doryman tried to fill until he had only a few inches of freeboard. Once the sea begins to come over the stern—always the deeper end—it is too late to throw out fish. A lifebelt would support only a frozen corpse.

After that incident, the size of the fish again showing an average decrease, though one taken by Mariano

THE GREENLAND CAMPAIGN—JUNE

da Cunha was 56 inches long, our little trio moved to the north once more, this time to the coastal bank south of Holsteinsborg Bay. This was on the Arctic Circle: we were to remain north of that for the following eighty days.

The fishing south of Holsteinsborg Bay was good but the weather was not, nor were the conditions. The bottom there was exceedingly bad, much broken by large pieces of rock which all along that coast had flung themselves to the surface, or just beneath the surface, where they lay in wait for unsuspecting ships. (The Danes were putting in a beam radio system, to assist coastal navigation in these dangerous waters.) The tides were bad, too, racing by at springs at anything up to five-and-a-half or six knots. Now and again there was fog, but not much as yet—just enough to show what that arch-enemy could do when he really set his mind to it. The very considerable magnetic variation and disturbances, sometimes causing the compass needles to swing badly in the boats, were an added danger. If a doryman should be seriously caught in fog, the only thing he could do was make for the fjords, with which the wild coast in those parts was deeply indented, and hope to achieve a safe landing. Long, murderous kelp waved in the water off much of that iron-bound coast, making access difficult, and the tides raced in the fjords.

In addition to the *Argus, Creoula,* and *Elisabeth,* only a few ships were on that particular shelf at the time. Fortunately, among these was the *Capitão Ferreira* with that excellent old sea-dog, Captain Antonio Marques, in command. Over sixty years of age, Captain Marques was one of the few masters in the fleet today who had himself been a doryman and had earned the right to command in the really hard way. The confident ring in his great voice on the many tense occasions when the captains consulted anxiously together—to launch, or not to launch; to recall, or not to recall—was an inspiration to them all.

The threats of sudden south wind were with us almost daily and, on too many occasions, more than the threats. Even if a threatened blow did not come to anything, it worried the dorymen. They did not hear the radio consultations. They only knew that they were out in their frail boats upon the Arctic sea, and they could read the signs. Fog and sudden wind—these were the greatest dangers. Though the fogs in June were neither so dense nor so clammy as those of the Grand Banks, they were still fogs and serious hazards. There were occasions when I began to fear for the dorymen. The fog so often shows mirages of the tiny dory sails to eyes which strain to see them, mirages which dissolve into nothingness as quickly as they appear. The siren wailed and the church bell tolled and at last the tiny triangular sails would really be there, coming through the fog, coming silently; and the dories would slip quietly alongside to gaff up their cargoes of silent dead cod.

"Boatswain! Salvador! Laurencinha!" Captain Adolfo would shout along the deck, after pencilling his estimate of the catches of the last dories. "The hands to dinner!"

The siren wailed no more for the moment and the church bell ceased to toll. Another day's fishing was over, though the cleaning and salting would go on late into the night. And, not far away, the wailing of another schooner could be heard, with all her dorymen not yet returned.

One such day, Monday, the 26th of June, we took almost three hundred quintals, and should have taken more if a freshening breeze had not forced the hoisting of the recall very early in the afternoon. There were more than a hundred quintals in the pounds by noon: before forty dories were back, the fish were spilling out of the pounds and overflowing on to the deck. The wooden working platform amidships was soon heaped high with fine, fat cod, and they poured into all the spaces between the wire-meshed pounds and over the tops of the refrigerator hatches. Half the night the loud-speaker was blaring music at a fast tempo, to help the work and encourage the men. Some of the deckboys were almost dead on their feet by midnight. One of them, from the Azores, was seated in a bloody pound, cutting out cods' tongues and throwing them into the sea, and putting the severed heads into the tongue basket by his side. It was grey, cold, and utterly cheerless, and the midnight daylight served only to emphasise the misery.

Not for the first time, I marvelled how the dorymen could keep going. No one brought round mugs of steaming coffee or hot tea, or indeed anything except a noggin of brandy about ten o'clock, though in all that long day they had eaten only a hurried breakfast, a cold bite in the dories, and one warm hasty meal when they returned, before the cleaning began. In these days, when the alleged need for making life easier has become a social postulate, men like our dorymen, inured to hardship and recking little of it, seemed an anachronism. They were unspoiled. They were men. There was work, and they did it. There were dangers, and they faced them. There were hardships and difficulties, and they overcame them. Endurance was ingrained in them: it had become part of their character.

I marvelled at the splendid manner in which our good dorymen always did their work and, again, at their courage and ability in the dories. Aboard their schooners, a merchant seaman might not regard them as outstanding mariners, for many of them obviously looked on the ship only as a dory-base, a sort of factory with a bunk-house attached. But in a dory! The sea that afternoon had run racing whitecaps all turbulent and threatening, for the wind was across the strong tidal race and making the water broken, rough, running far from true. Almost every dory was full, even the Little King's: north of the Arctic Circle his ancient spirit blossomed and he toiled manfully—some days— to fill his dory. They had gone for miles. Hours after the recall had been hoisted, the tiny pin points of the triangular sails were still breaking the windy horizon as the little red dories raced towards the mother-ship. They were not all to wind'ard. Some poor wretches had to beat, for the ship could not shift to a satisfactory position alee for all of them. All round, the sea boiled furiously, angry at the dories passing, yet curling back from the gunwales of the flying craft, as if the wraiths of all the drowned Greenland dorymen were racing there with them, holding back the sea. Each dory had a roll of white water at the sharp bows and white water everywhere, the little masts bent like whips (for they are only sticks cut from young trees, back home in Portugal), and the only rigging the single-strand halliard of the minute sail, set up taut and secured to the rail to windward. The sea through which they sailed was growing ugly and looked unlivable even for a good lifeboat. Yet a dory is nothing but a little pointed box, heavily flared, an affair of nailed planks, undecked, with neither centerboard nor keel nor even a rudder, designed to carry fish and to be nested, without buoyancy tanks, overloaded—if the doryman can possibly contrive it so—to an extent which would make it dangerous to handle in a river, or upon the waters of a quiet protected bay. This was the open sea, within the Arctic Circle, and the wild water raced over treacherous banks which seemed designed to rouse the anger of the restless sea. To be spilled into that water was certain death. Many dorymen had so died, and it was easy to imagine their ghosts now racing there and in the fogs, with their decrepit, weather-worn dories and their threadbare sails. Perhaps these were the sails we were for ever seeing in the snow-squalls and the fogs, when the living dorymen did not come—threadbare sails from the bottom of the sea.

As each little fellow came silently and speedily to wind'ard of the pitching ship, he rounded-to smartly and let his sail down on the run. In the troughs it was often almost impossible to see him even when he was no more than a few fathoms from the ship. Then he manoeuvred alongside with infinite patience and care, watching the run of every sea and the spray from all the breaking water. He approached the ships under oars alone, for he realised that this was the really dangerous part of his returning. He threw out the bloody water furiously with his scoop bailer, watched his chance, swung head to sea, dropped alongside. A deckboy, sleepy and cold, picked up the bight of the dory's painter on a long hooked pole and hauled it quickly inboard. Another passed the dorymen a gaff, and he began at once to gaff up his load of fish with energy and haste, for he was well aware that his life might depend on it. When he came aboard at last, he clumped in his heavy sea-booted feet, a burly, ungainly giant in his swathes of oilskins. Ungainly he might seem to be, but he was one of the best small-boat sailors in the world.

The days passed slowly. Our loss of gear—grapnels, and sections of long-line—was serious. But the fishing continued to be reasonably good and we recovered all our dorymen every day. By the end of June, César calculated that compared with the corresponding day the year before we were a thousand quintals ahead. Even according to Captain Adolfo's estimate, we had three thousand quintals stowed, and that was a full thousand better than in 1949. In 1949 she had not taken a capacity cargo; but we felt slightly encouraged. Once or twice, I thought I saw a smile flit across Captain Adolfo's dark face, when a dozen or more deep dories were racing for us in good weather by mid-morning, and the horizon was unbroken by any little sails in quest of better ground.

CHAPTER ELEVEN

FIRST FISHER OF PORTUGAL

Brave in action, patient in long Toyle.

"I SOMETIMES think they must be different from other men," said Commander Tavares de Almeida, in the *Gil Eanes,* as he watched the rescue in a dreadful fog of a doryman who had been adrift two days. It was not really a rescue. The doryman had not admitted to himself that he was lost, and he had found the *Gil Eanes* by chance. He climbed up her high pilot-ladder without assistance, and his fish and then his dory were hoisted aboard until his schooner could be found. His face darkened by exposure, his hands bloody, cracked, and swollen, his back bent as if by the weight of years though he was a man in the prime of life, he stood there impassive, unperturbed, asking for nothing. He had made no signal. He had been engaged, he said, on a square search for his schooner. In answer to questions, he said he thought she was to leeward somewhere. The weather had been bad the previous day: he had put out his anchor at night, and sheltered under his sail. That was all we ever heard from him: the next day, he was back in his schooner and fishing again.

What kind of men were these Portuguese Arctic dorymen? Old Antonio Rodrigues, in the *Argus* on his forty-second campaign; old Manuel de Sousa, who grew thinner and darker and more haggard every day, yet always fished as well as many a younger man; that fine old gentleman, Jacinto Martins, Jr., who was foreman splitter of the Azoreans; and the younger men, like our redoubtable First Fisher and the First Fisher of the Azores? What was their background? What were their philosophies, their hopes, their plans? What did they do for the rest of the year, when they were not driving over-loaded dories dangerously in the Arctic seas? I tried to find out. I spent much time with these men, yarning down in the rancho on bad-weather days when there could be no fishing, and sometimes out in the dories. They spoke only Portuguese and, some of them, a strange Portuguese. Even the mates had difficulty in understanding some of them. Portuguese is a difficult language to understand when spoken rapidly with syllables clipped almost to extinction; but I found I could understand the dorymen better than the afterguard. They used fewer words. The Azoreans had a curious accent of their own and this often defeated me; but there were always interpreters to help me in my inevitable difficulties. As the voyage progressed all the afterguard improved their English, and the two mates were very good. So, in the end, I became able to understand what was being said and, as the weeks and months slipped slowly by, gradually got some picture of the dorymen from themselves.

Consider, for example, such a man as Francisco Emilio Battista, called Laurencinha (a family name), First Fisher of the *Argus* since she was built, of Portugal for a decade, and more likely than not, of the world. He was a lithe, slight man, without an ounce of superfluous fat, with a strong, dark face, and fierce, imperious eyes. He had strong wrists, great hands and a ready smile despite the hardships of his chosen life. Above all he had an infinite capacity for thoroughness in all he did or thought of doing, singleness of purpose and the ability to go straight for what he aimed at, and to keep going despite all difficulties. His energy was boundless, and his strength of will indomitable. It was difficult to understand why he should have possessed all these qualities to such a striking degree. His brother Leandro, in the *Argus* with him, was a good fisher, but not outstanding. His brother Manuel, in the *Inacio Cunha*, was also good, but not First Fisher. True, his father José, who fished for forty campaigns upon the Banks and died in the Algarve at the age of seventy, had been First Fisher for many years in the schooners in which he sailed. His nephew José, in the *Argus* with him, was a first-class doryman and an excellent seaman —the two qualities are by no means always found together—though he was only twenty-two and had made but five voyages.

Our First Fisher was born at Fuzeta, in that province of Portugal which is still called the kingdom of the Algarve, on the fourth of August, 1914, and his father and uncles were away on the Banks at the time. He was the youngest of four brothers, all destined to become Arctic dorymen. He had no schooling. None was available for fishermen's sons when he was a boy, though it is now; and his nephew, the young José, went to school. Even if there had been a school, it is doubtful whether he would have attended it regularly. Fuzeta men preferred the school of life. This was provided for him abundantly. Before he could walk, he had been out in the dories and the small sailing-vessels, *caiques* and *buques,* which went out fishing beyond the bar. He could sail a dory as soon as he could walk, and he was a fully fledged, and paid, hand in a small *caique* when he was eight. When his father and uncles came back from the Banks they always went fishing in the small sailing vessels from Fuzeta. A boy was considered old enough to be useful when he was eight. A *caique*— a type in common use then but now almost gone—was a swift and narrow craft, lateen-rigged, splendidly seaworthy but allowing no fool to handle her, just the sort of little ship to develop brightness and the qualities of good seamanship in a lively boy. The youthful Laurencinha did well, and soon made a name for himself. He delighted in the life, especially when the Bankers were home, and the *caique's* crew was brought to its full strength of twenty-five. In the summers, when so many men were away, she sailed with only fifteen. The Fuzeta men went to the Banks because they had been going to the Banks for generation after generation. Fuzeta dorymen had long enjoyed a reputation for being pertinacious and fearless, and the Banks captains liked to recruit from the pretty fishing port. Laurencinha was marked out as a doryman from the day of his birth.

For five years, learning all he could, he sailed in

caiques, buques, and dories. A *buque* was lateen-rigged, like a *caique,* but stouter and blunter, and seldom with the same grace. A small Algarvian *caique* had once sailed to Rio de Janeiro.

This had happened in 1807, at the beginning of the Peninsular War, when the Algarvians were among the first to fight. This they did—against heavy odds—with such success at Olhão (in the Fuzeta district) that the townspeople decided to send the good news to the Prince Regent at Rio. But how could this be done? There were only small fishing craft left in the port. Two stalwart fishermen, by name Manuel Martins Garrocho and Manuel de Oliveira Nobre, offered to sail a small *caique* out, and this they did, to the astonishment of the citizens of Rio where they were given a tremendous reception. They were granted rank in the navy, given pensions, and the *caique* itself was preserved ashore in the Brazilian capital while the two mariners returned to Olhão in a new vessel, which was built for them. The *caique* may still be at Rio. In earlier days, larger *caiques* had sailed to the Banks.

Buques were employed in the coasting trade, and fishing. They were good little ships, and still are, and so were the other local type, the single-masted *canoas* which carried one huge lateen main. The boy's experience in such vessels was excellent preparation for the future Banker. He kept asking his father when he could go with him upon one of these long voyages of which he heard so much. At nearby Olhão, at Faro, at all the lovely Algarvian ports where the little ships touched to sell their fish, the talk was always of great adventures on the distant Banks, of hazards faced and overcome, of famous captains and their deeds, and of the miraculous escapes of fortunate dorymen. An uncle, lost in a fog somewhere on the Banquereau, had turned up in Fuzeta years later, when he had long been mourned for dead. He had been picked up by a big sailing-ship after being adrift four days. The weather was bad, the sailing-ship did not intend to lose time on an already long voyage by looking for a fishing schooner in a Banks fog, and she was bound for China. To China the doryman had to go, and it took him some years to get back again. Francisco had heard his uncle tell of this adventure when the fishermen, unable to fish because of bad weather, gathered round the stone quay at Portimão or the front at Olhão, or by the *buques* in the basin at Faro. He had no particular desire to be carried off to China but, to a boy who all his life had known nothing but sailing-ships, the prospect was not alarming.

When he was thirteen years old, his father took him as deckboy in the three-masted schooner *Rio Lima*. His father was First Fisher of the schooner at that time. In those days there were no such things as auxiliary engines or even power windlasses. The *Rio Lima* simply sailed from Lisbon, went to the Banks, fished there for five months, and then sailed directly back to Lisbon. It was a bad season, with much bad weather and the cod far from plentiful. The schooner fished less than half a cargo, and his life as a deckboy did not impress the small boy favourably. It was not his idea to slave the long days in the humdrum salt, or be shouted at, day in day out, for all the trifling jobs. The dog's life of a Banker's deckboy was not for him. Deckboys were not allowed to go in the dories, and he soon realised that he would learn more about fishing by con-

In the Straits of Belle Isle

Away Went the Dories Once More

A Dory Swamped

Full Dory

The First Fisher's Dory Was Full Almost Every Day

tinuing in the local craft at Fuzeta. So from the voyage in the *Rio Lima* young Laurencinha went straight home, and stayed there for the following five years. In those days, dorymen usually began at eighteen or nineteen years of age. Theirs was no life for a boy.

At eighteen, Laurencinha went back to the Bankers, this time as a doryman in his own right. His father was an old man then, and dying. His elder brother José was Second Fisher of the beautiful *Hortense,* flagship of the Bensaude fleet. It was customary for Fuzeta men serving as dorymen to recommend relatives and friends to their captains when opportunity occurred, and this was usually the way in which a young fellow got his start. José Battista recommended his brother Francisco to Captain Anibal Ramalheira of the *Hortense,* a leading figure in modern Banks fishing, and a word from José was enough. The young man was signed on as green doryman. At the end of his first season—again, a tough one—he ranked fourteenth of the thirty-eight fishers aboard, and Captain Anibal had already marked him as an unusually promising doryman.

He made two more voyages in the *Hortense.* On the second, he ranked fifth. On the third, he passed his brother to become First Fisher of the *Hortense,* and he has remained a First Fisher ever since. In the earlier days there were no central records, but when the Gremio came into being and kept full records of the industry, it soon appeared that the *Hortense's* new First Fisher was also the best man at his business in the whole fleet, and that easily. This record he had maintained, except for one year when an accident with a knife, when he was splitting, caused him to lose several days. Even then, he was only two quintals behind.

At the time the young man showed his worth, the enterprising Vasco Bensaude, always a pioneer in long-voyage fishing, had decided to build a new kind of Banker, the first of the modern steel schooners, with refrigerated bait, echo-sounding, ample auxiliary power, a full sail plan, and other improvements. This was to be the *Creoula,* to replace an earlier vessel of the same name, and besides pioneering with the ship's equipment, she was to try out the Canadian-type (and Algarvian) long-line, for the first time in that fleet. Captain Anibal Ramalheira was appointed to command, and he took his best fishers with him from the *Hortense.* At the head of these went Laurencinha. The *Creoula,* though inclined to be wet, was a success from the outset, and so was the long-line. So also was Laurencinha. He stayed three voyages in the *Creoula.* On the first, she was late from the yards and did not sail from Lisbon until June 29. For this reason she did not take a full quantity of salt, since there seemed no hope of fishing a complete cargo or anything like it. But in fact she could have done so. She had used up all the salt within three months and was back in Lisbon a fortnight later, from Greenland.

"That was the shortest voyage I ever made," Laurencinha said. "The average, in these big ships, has been five-and-a-half to six-and-a-half months. They take too much fish, and they always grow bigger the more fish you put in them."

Meanwhile the *Argus* was being built in Holland, to be queen of them all, and Captain Anibal was standing by. Captain Adolfo took the *Creoula* on the last voyage which Laurencinha made in her but, as soon as the new *Argus* was commissioned, he transferred with his old captain to her, and with her he has been ever since, throughout her twelve voyages. First with the brothers Ramalheira, Captain João—called Vitorino—who succeeded Captain Anibal; and then, for eight voyages, with Captain Adolfo. He has not yet been wrecked, or foundered, or sunk a dory; nor has he had to swim for any reason on the Banks, nor been lost overnight in fog. That, in itself, is a remarkable record. He has had the good fortune to serve with excellent captains in outstanding ships, each in her turn the pride of a well-run and successful fleet. But his way was not made easy and it is not easy now. A remarkable exponent of a dangerous and exhausting craft, his success is entirely due to himself.

I watched him, for once near the ship, on the day after our long yarn. There was a threat of southerly again, and he had shifted closer to the ship to make his third shoot for the day, to be safer if the sea got up. Some dorymen were not making a third shoot at all, and others had shortened their lines. Other dorymen hauled in their lines slowly, taking a little time over each fish. The First Fisher always hauled in his fish boldly, hand over hand, as rapidly as he could, no matter how the dory was jumping, flinging them with a flick of his elbow into the dory and, with a second rapid, practised flick, coiling down line, snood, and hook clear in the tub. One look of his fierce brown eyes was enough to tell him whether big cod were securely hooked in their mouths, or held only by a prick of their rubbery lips or by the loose skin of their cow-like faces. He gaffed them in when necessary, losing no more than a few seconds, and so he continued until the whole long-line was in, without pause, though it must have been a great strain on his arms. Then at once, setting his little sail, he sailed the line out again, all hooks clear, rebaiting quickly with a piece of catfish or fresh halibut as the unbaited hooks came to the top. As soon as the line was shot again, he set to work jigging, a jig-line in each hand. What a difference between this performance and the Little King's! Even though the King was trying, and trying hard.

Watching him at close quarters I could understand why the First Fisher excelled: usually he went many miles from the ship. I asked him how it was that he could always fill his dory. But he knew no answer except work. It was more than that. Work, indeed, but work, and a clear head; work, and magnificent hands, with the true fisherman's sensitive fingers to receive all the messages flashed lightly from thirty, forty, even fifty fathoms down; work, and the indefatigable determination of a strong-minded, magnificently fit man, who would have been a success at anything. And, as a background, centuries of the same tradition. No one knew how long the Fuzeta-men had been fishing, possibly for thousands of years. And behind him, too, was the sure shield of a great religion, a sure belief in God. Francisco Emilio Battista was a simple man, an illiterate, and consequently debarred from entry to the U.S.A. and my own land of Australia. Yet he was a man whose knowledge was his own, gained and secure in his own mind, a man educated by life and not misled by nonsense, a man who knew and got on with the job for which he was fitted, a man untroubled by political slogans, uninhibited by any confusion of besetting doubts, unhampered by theories, complacent or destructive. His way of life was set, and for him and his happy kind it was enough: his feet were upon the earth though half the year they trod the frail planks of an Arctic dory. He was a man, I thought, very close to God—he, and all the dorymen. Down upon the sea, his life in his hands, he knew the way of the sea and the way of the Lord. And, knowing these things, he was fortunate.

He had six children, four boys and two girls. Four were already at school; all would go. His ambition was to save enough to buy a *buque* or a *canoa* of his own, preferably with an engine, to fish from Fuzeta all the year round, and to come to the Banks no more. When he was not thinking of his work, sometimes quick thoughts of wife and home and children came into his mind. Like all the dorymen, he was much attached to his distant home and family. He hoped the day was not too distant when he might buy his *buque;* a man past forty ought not to be an Arctic doryman in a big hand-liner. But for the time being, he was happy.

So, too, were Antonio Rodrigues, Manuel de Sousa, old Jacinto Martins and Francisco of the same name, First Fisher of the Azores. These were all dorymen born and bred, as almost all the others were. Each had been in the *Argus* since she was built, chosen by Captain Anibal to serve in that fine ship. So were João de Oliveira, the Second Fisher, and the Martins brothers from the Algarve, Salvador and the Star. The brothers Ramalheira and Captain Adolfo knew how to choose their men.

João de Oliveira was another great doryman who had been "discovered" by Captain Anibal. All his experience on the Banks had been gained in the *Hortense* and *Argus*. In 1950 he was thirty-one years of age. He had once tried his luck on the West African fishing grounds, out of Luanda, but he said the Arctic was better than the tropics and he wanted no more of Luanda. Though in the *Argus* he was overshadowed by Laurencinha's extraordinary skill, João de Oliveira was an excellent doryman and would have been First Fisher in his own right in almost any other vessel. He had two children, had been four years at school, and was a member of a very old fishing family which had been sending dorymen to the Banks as long as Portuguese dorymen had gone there. His brother Sebastian, aged twenty-three, had been four years in the *Argus* and was already among the twelve best dorymen—in that ship, an honour indeed. Sebastian was a cheerful young man, always ready with a smile, but João was much given to reflection. He had had to swim more than once, and had narrowly escaped with his life. At home in Fuzeta he owned a fishing-boat, and his ambition was to replace her with a larger, as soon as possible. She was only twenty feet long and could not make a living for a family the year round. In the *buques,* he was a "master of the nets," a master-fisherman, and he was accorded some respect by the others on account of this.

The slight, dark figure of ancient Antonio Rodrigues was striking, his strong features darker than ever from months of exposure, except where he had spilled some crimson antiseptic on his chapped and wrinkled skin to defeat infection. His grizzled grey beard was squared like an Egyptian fisherman's, his big hands wrinkled and white and torn. They, too, were covered with crimson antiseptic. He never missed a day's fishing or fell out for a moment from the cleaning line. Always dressed in several brightly checked shirts with a giant waistcoat of pilot-cloth which left his arms free to fish, checked trousers, thigh rubber sea boots—the company's—old Antonio had been a doryman since 1907. His father had been a fisherman; his two sons were in the *Hortense*, one of them as First Fisher there. The old man had been a First Fisher in his time, but he was past it now. He had been shipwrecked, sunk in fog, adrift in his dory. He knew the hard old days when the doryman's life was much more difficult than it is now.

"I wish I were just beginning!" said the old man. "It is all right now. Engines, electric light, plenty of bait, a night's sleep now and again, dry bunks, an assistance-ship on the Banks with mail from home, a radio programme for us, going into St. John's or North Sydney to break the monotony—aye, it's a good life now!"

He had always liked it, he said, even at its worst. When he began, most of the modern improvements now taken for granted had not been thought of. Conditions in some of the old ranchos were indescribable, though the old-timers who lived in them did not seem to notice. It was a strange thing, but neither the dorymen nor the owners seemed to want the Gremio when it was first organised. It interfered with their "freedom," they said. Antonio himself had taken to the hills of the Algarve for a time in a mistaken attempt to preserve what he imagined to be his freedom. He was First Fisher of the old *Argus* then. After a while, the dorymen were rounded up and sent back to their ships. Since then, there had been no trouble. He had been nineteen days in the hills. The old man, his deep sunken eyes sparkling at the recollection, grinned. It must have been a queer adventure, with the dorymen from the

Argus and a dozen other schooners turned gypsies in the hills.

"It was tough, but we enjoyed it," he chuckled. He had been forgiven long ago, and now possessed a special medal, pinned on him by President Carmona, for his great record as a doryman on more than two-score voyages. The medal was at home, where sometimes he took it from its case and looked at it, and showed it to his six surviving children and his six grandchildren. He still fished regularly from Fuzeta during the winter months.

Manuel de Sousa—called Vinhas, the Vines, to distinguish him from the many other de Sousas—was another elderly doryman. Grey-haired, with fine features and kindly dark eyes, he was slight in build, a sailor as well as a doryman, and on the articles of the *Argus* in both capacities. He had been a sailor and a doryman for more than forty years, though he was only fifty-five. Schooners, barquentines, brigantines—he had served them all, but never a steamer. His father was a Fuzeta doryman before him, and his elder son was with him in the *Argus*. His own elder brother had first taken him to the Banks in a little old wooden barquentine called the *Terra Nova*. His many ships included the *Maria da Gloria*, with Captain Silvio Ramalheira; he had been a good deal with the Ramalheira's and the Paião's, as most of the crew had been. He had lost a dory once, but fortunately it was near the Virgin Rocks on the Grand Banks, in summer: round the Virgin Rocks there were many other dorymen at the time, and he was picked up. As a sailor, his winter employment was to assist in the maintenance of the *Argus* at her lay-up berth. His wife usually came up from the Algarve to join him and lived aboard in a sort of compartment which they contrived in the rancho. His son, who had been a doryman since he was eighteen and was now twenty-six, fished through the winters. From the lay-up berth to his home in Fuzeta was a long overnight journey by train and so he did not often go home. His wife was able to live aboard until a week or two before sailing. Their children were all grown up and married.

Old Manuel de Sousa was cousin to the captain from the Algarve in the *Santa Isabel*. Force of character, unusual ability, and an American background—he had fished out of Gloucester for years—had helped to bring advancement to this cousin. It was unusual for a doryman to have a close relative in command, though the cook and the captain of one of the motorships were brothers. A cook, however, was rated as a highly skilled and important key man.

Jacinto Martins was from the Azores. Like the majority of the Fuzeta dorymen the Azoreans usually went to sea as competent fishermen, not as deckboys. They were fishermen first and Bankers second though, in almost all cases, there was a long tradition of Banks voyaging in their families. Old Jacinto—he was fifty-nine—had been making Banks voyages since 1909, but he had been fishing since 1900 from his home town of Ponta Delgada. In the winters, he still fished from Ponta Delgada where he had his own small boat, about eighteen feet long. He had been in some queer little Bankers in his time. Before the 1914-18 war, he said, vessels of less than 100 tons used to sail from Portugal, and some of them could carry only twelve or fourteen dories. They always considered themselves very lucky if they brought all the dories back again. The sea took them, nearly al-

ways on the return voyages. They were tough little
ships. But now, said Jacinto, the doryman's was a peaceful life, with a fine safe ship to live in, the bait in a
thirty-ton refrigerator, and a dry bunk in which to sleep.
Aye, a good life—now; and down there on the sea in his
dory, a man could fish peacefully all day, his own master. But his ambition was to settle down at Ponta Delgada and to be an Arctic doryman no more. At sixty,
he thought, it was nearly time to retire. Hauling the
long-line in deep and fiercely tidal waters was work for
young men, and the Arctic season could be long and
hard.

Francisco Martins, the First Fisher of the Azores, was
no relation to Jacinto, though he had a brother aboard.
He was a handsome young man, a born fisherman, and
all his Banks experience had been in the *Argus*. He
joined her for her maiden voyage and had made every
voyage since. His father was a Banker before him: fishing was the only life the family knew. He had three
sons, the eldest of whom was nine years old. They were
going to school, and he hoped they would not be dorymen. He had had to swim several times, and he knew
he had been lucky to survive. He did not want to expose his children to those risks, though he accepted
them cheerfully himself. He would like them to go to
America, or perhaps South Africa, or Australia, for there
were too many people in the Azores. It was not good to
remain crowded upon an island, even so lovely an island
as St. Michael's. As for himself, he was happy enough
at his work—he certainly always gave that impression—
and he liked the *Argus*. But six months was a long time
to be gone from home.

Francisco Martins was an energetic and skilful doryman with abilities far above the average. His dory, number 51, was filled almost every day, often twice a day. Some weeks, he was Second Fisher in the ship, passing João de Oliveira's total, and there was considerable rivalry between them. Francisco had to fight for his place, too, against several of his fellow islanders, particularly the footballer Raul Pereira and the youthful doryman Manuel dos Santos Rafael. Like himself, these were fishermen born and dorymen bred, robust and indefatigable experts who filled their dories with fat cod consistently, day after day. So did César de Medeiros, a black bearded young man on his twelfth voyage who looked like a pirate and fished like a one-man trawler. They were all the best of friends, though fierce individualists in their dories. At cleaning and salting fish they worked splendidly as a team, but they were one and all decided in their preference for the one-man dory.

"No two men are alike," Laurencinha said.

"Aye, aye, the two-man dory might be safer, I'll grant that," said Jacinto Martins. "But we're used to this. We'll keep it this way."

"My dory is my sea-horse," said Francisco Martins. "I don't want any rump-riders."

"I will catch my fish myself," said João de Oliveira.

"There have been experiments with two-man dories," old Antonio Rodrigues recalled. "They make sense. I have seen a good many new ideas tried in my time and they have nearly all been good. Maybe the two-man dory is good, too, but we are accustomed to fishing alone and we'd like to see things stay that way."

The truth was that that they were as conservative as they were courageous. New ideas in the *ship*—yes: they were acceptable and the dorymen could see their great advantages. But new ideas in the dories were entirely different, and not to be tolerated. The ship was something in which they lived and slept and cleaned and salted fish; but their dories—why, they were their lives.

CHAPTER TWELVE

THE CAPTAINS FROM ILHAVO

The valiant Portingalls that plow the Main.

FOR the thirtieth time that morning, Captain Silvio Ramalheira of the motor-ship *Elisabeth* paced impatiently the weather wing of his minute bridge, his blue eyes roving intently over the wide seascape, from the dark mass of the West Greenland shore way to the eastward, round the assembled schooners of the Portuguese hand-lining fleet, over the sea and sky. What was the weather going to do? That was the problem. That was always the problem; and a man could guess no better at the answer, here in Davis Straits, after a dozen or more voyages. He had launched his fifty-five dories a little after four that morning, as all the other captains had done, doubtfully indeed, for the weather was far from settled. But if they kept the dories nested whenever there was doubt, they would never fill with fish. The dorymen had gone cheerfully.

THE QUEST OF THE SCHOONER ARGUS

The dorymen went at the captain's bidding: their lives were in his hands. He sent them; he had to get them back. Captain Silvio had an excellent crowd of dorymen and, like all the other captains, he felt very keenly the responsibility for their lives. Like most of the other captains, too, he had lost some lives. It was no fault of his but it distressed him profoundly. His eyes fell upon the white speck of a small sailing-ship with square yards, far off by the Greenland mountains. He knew that speck well. That was the barquentine *Gazela*. Captain Silvio had been a deckboy in her on his first Grand Banks voyage almost forty years before, though he was still a man on the right side of fifty. He had been mate in the *Gazela* and had been several years in command of her, later. In her, he remembered—he could never forget—he had lost two dorymen.

He was fishing that day in July, 1932, off Holsteinsborg Bay, in the treacherous currents there, where a master had to be doubly careful. A sudden southerly wind, springing out of a clear sky on what had been a lovely day of good weather and good fishing, brought up a breaking sea. The *Gazela* had not then made many Greenland voyages, and neither had her youthful master. The barquentine had no power in those days: he could not weigh at once when he saw bad weather approaching, or when it was suddenly upon him, as it was that day, without giving any sign of its approach. He could not drift down among his dorymen to pick them up and save them the dangerous battle back to wind'ard to regain the ship. He could do nothing. The *Gazela* was securely anchored to a long scope of cable and there she had to stay, for it took at least four hours for all hands to weigh her great Banks anchor, and she had only the cook and a few deckboys aboard. It was not in Captain Silvio's character to do nothing. In those days, if the ship herself could not reach her dorymen, it was the custom to float back a long grass line to them, over the stern of the anchored ship. This line was generally buoyed on an empty dory. The dory was launched, the grass line attached, and away drove the dory astern carried by wind and tide and taking the line with it, out to its utmost length. Then the dorymen made for the grass line and, once with a turn of their painters on that, they could be hauled up to the ship. It was primitive, but it was all that could be done and it was usually effective.

Captain Silvio went in the dory himself that day, taking the line. He took a second dory too, and in this he rowed and sailed to his distressed dories until he had found and succoured all that he could see, towing the heavily laden, encouraging the disheartened. He had thirty-one dories out that day. Twenty-nine he found: the other two never came. The wind screamed at the little barquentine, roaring in her square rigging, and the sea leaped at her exultantly. Hour after hour, Captain Silvio in his dory fought wind and sea to save his fishermen. But the lost two did not come. A deckboy said afterwards that he had seen one overturn near the ship, quite close by; but there was no other dory there just then, and the doryman did not rise again. Next day, when the *Gazela* was miles from there, that dory returned to her and gently bumped her stem; and in the afternoon, the lost doryman's dinner-pail drifted slowly by.

Many dorymen were drowned that day, from many ships. The dangers of the sudden southerlies were not then fully appreciated. They had been, ever since. Was the infernal weather going to do the same sort of thing again? It might. The only certainty, indeed, was that sooner or later, it would. But when? Captain Silvio knew he had ample power under his sea-booted feet now. Gone were the days of drifting back grass lines. He could have warning of southerlies, too, from captains further to the south than himself. But he needed good warning. Fifty-five dories, each two miles and more from the ship, each with at least a 600-hook long-line down in twenty fathoms of turbulent water, needed time to recover their gear, return to the ship, gaff up their fish into the pounds, and be hoisted aboard. He looked at the narrow platform across the main deck on the fore part of his bridge, where the six nests of dories had to be stacked: he looked and he hated it, for he knew that the curse of the motor-ships was that they could not always recover their dories quickly in a sudden blow.

Away on the southern horizon, he could see the four masts of a big steel schooner and, not so far away, another of the same sort. These were the *Argus* and the *Creoula,* and their captains, the brothers Paião, would be as worried about the weather and their dorymen as he was. He knew that. He had spoken to them both, over his microphone, six times that morning already. Adolfo in the *Argus* was not always strictly accurate in his reports about his fish; but he was a good fisherman, an experienced Greenlands campaigner, and a good weather-guesser. So was his younger brother. They were still fishing and did not think of hoisting the recall just yet. But they could hoist dories inboard all along their low main decks. If necessary, they could sail them in-

THE CAPTAINS FROM ILHAVO

board, take them over the side as the schooners rolled. Motor-ships were short and stumpy compared with schooners. The schooners were long and lean, to sail well: the motor-ship needed no such provision. So her short main-deck had room only for her pounds and the fish-cleaning gear, and the dories had to be stacked on a false deck above, and wherever they could be fitted round the funnel and the engine-casing.

There was much to be said for the schooner. Captain Silvio knew this well, for he had been in many of them and had been part-owner of his last. He had been years in the old tops'l schooner *Creoula,* forerunner of the graceful four-master which now bore that name, in the swift *Neptuno* and the *Gamo* as well as in the *Gazela.* And the *Maria da Gloria*—ah, the *Maria da Gloria!* He had not been in a schooner since the U-boat had shelled her. He hated to think of it, even now. The *Maria da Gloria* was a lovely three-master, a wooden schooner, graceful and able. His life savings were in her, and he loved that ship. During the 1939-45 war, she continued to fish on the Grand Banks and in Davis Straits. One day, out of a clear sky, shells suddenly began to burst round her, coming apparently from nowhere. He had thought at first that he had failed to notice some grey wolf of a warship signalling him to heave-to for the usual wartime identification, a Britisher, or a Canadian, to check that he was no Graf Felix von Luckner, and the *Maria da Gloria* no new *Seeadler,* or lesser *Graf Spee* in disguise. But the horizon disclosed no warship, though the visibility was good.

Then the shells began to burst aboard. Several dorymen were killed. A shell struck the jib-boom, bringing down the foremast. Some nests of dories began to burn.

Some of the shells were "anti-personnel," designed to kill his people rather than to damage the ship. She, unfortunately, was already damaged enough, afire and sinking. He gave the order to abandon ship. Nine dories got away with the survivors. Then he saw what was attacking him. A big submarine, which had been shelling from the surface, came closer to finish off the *Maria da Gloria*, and did so. They clearly saw the German submarine which, when it had sunk his schooner, turned its attention to the surviving dories and began to shell them. A shell burst in the middle of the nine dories, and Captain Silvio was among those gravely wounded by its flying fragments. The submarine then went away. Almost immediately, the wind and sea got up, from the southwest. Within an hour it was blowing a gale. The dories were secured together and tried to ride to sea anchors, but some had their gear, and others had not, for they had had to be launched quickly from the nests. Since the schooner was on passage to new grounds at the time they were not all rigged for fishing. The weather and the high sea caused some of the dories to break adrift and these were not seen again. Many men were badly wounded. Some died. Some went mad from their sufferings. They had no food and no water, and the gale blew for four days. During this time Captain Silvio, just sufficiently conscious to keep control, shaped a course towards the coast of Labrador or northern Newfoundland, crossing the tracks of the Greenland-bound schooners which, he hoped against hope, might pick them up. But they saw no schooners.

On the fifth day, only three dories remained. Meanwhile, Captain Silvio's condition was rapidly worsening. His dorymen thought it best to put him in the mate's

THE CAPTAINS FROM ILHAVO

dory, which was still in company, and so they did. That night, they were lost, and only the mate's dory was left. After nine days, an aircraft on anti-submarine patrol sighted them and dropped canisters of flares and food. Two days later, these flares lighted the American auxiliary *Sea Cloud* to the last dory. There were only six survivors, and it was many months before Captain Silvio could walk. When he could, he went to sea again. Thirty-six men had died with the *Maria da Gloria*.

Captain Silvio did not know it at the time, but it was the misfortune of his schooner that she was across the probable track of the fleeing *Bismarck*, and so she had to go. He did not see the *Bismarck*, and did not even know that the big German had broken out.

He shuddered now as he thought of it, and his scars hurt. They always hurt in Davis Straits. The doctor said it was imagination, but the doctor had not been eleven days adrift in a dory.

Captain Silvio grasped his microphone, threw the switch, and began to call young Captain Leite in the *Gazela*, nearer inshore, and Captain Adolfo in the *Argus*.

"*Elisabeth* calling *Gazela* and *Argus*," he began, in his strong voice. "*Elisabeth* calling *Gazela* and *Argus*. *Elisabeth* calling . . ." From the microphone on the bridge, he could see out over the fore-deck where his cousin the mate, Elmano Ramalheira, was superintending the work. There were many Ramalheiras of Ilhavo with the codfishing fleet. Another cousin, Manuel, was in command of the schooner *Infante de Sagres*. Yet another cousin, João Pereira Ramalheira called Vitorino, had the hospitalship *Gil Eanes*, which should soon be arriving. Anibal Ramalheira, brother to Vitorino and

Elmano, and Silvio's cousin, was marine superintendent for the Bensaude line ashore at Lisbon, after a long and distinguished career in command of the schooners. It was this Anibal who had first tried out the revolutionary *Argus* and *Creoula* and who, with the dynamic Vasco Bensaude to back him, was largely responsible for many of the innovations which had been made. Radio telephony, for example: only a few years before, the old die-hards of the last generation of Ilhavo Banks masters had been sworn opponents of the very idea. What! Talk about fishing over the free air, for all men to listen? Tell the fleet when you were on fish, and bring the hungry lot of them to spoil the ground? Not likely! They had managed a few hundred years without such new-fangled things as pieces of black plastic called microphones to help them to bridge the gap across the miles of sea. They did not at first realise the life-saving value of the innovation. But the Bensaudes and the Ramalheiras went ahead. They had gone ahead, too, with the long-line, with echo-sounding, with contracts for the supply of refrigerated bait. And they had supported the Gremio in many of the ideas which owners, captains, and dorymen alike opposed, excellent as the ideas later turned out to be.

How long the Ramalheiras have been going to the Banks no man can say with certainty, for there are no early records of Banking crews at Ilhavo. Many a ship hailing from the neighboring port of Aveiro had been manned by mariners and masters from Ilhavo. Ilhavo has almost a monopoly in providing masters for the Banking schooners; and the Ramalheiras have been in the trade as long as any. The men of Ilhavo have always followed the sea. There is little choice for them, indeed, for the arable land has long been parcelled out until most inheritances mean nothing. A man can not hope for land enough to provide his family with a living. The arms of the sea running inland from Aveiro, and the open Atlantic so near, provide a perfect training ground for boys destined for the sea. To this day craft strongly reminiscent of the slow-moving, lovely vessels of the ancient Phoenicians move quietly and with grace through the waters of the estuary, still propelled by sails. On the Douro, not far away, the wine boats make their seasonal passages under sail. The sail is still accepted there as the ideal means of imparting movement to anything waterborne.

The mariners of Ilhavo sailed many a deep sea schooner and square-rigged ship, until the day of such ships was done. Now they sail about their estuary in their distinctive small craft; in coastal and Mediterranean schooners, in the Grand Banks and Greenland fleet. The Monicas at Gafanha, their yards in sight of the Aveiro bar, built lovely schooners, brigantines, and barquentines for generations. There were others before them. Ilhavo is a town of ships and shipping—almost the last place in Europe where a barquentine could be manned for a voyage to the Banks or a run to Rio; where a brig could be fitted out for a treasure hunt to the Cocos Islands, and sailed there. Sail-makers, shipwrights of the old school, rope makers, builders of schooners and of dories, all flourish in the area, their skills now centered on the Arctic sailing fleet. Not only the great majority of the schooner captains but also a large proportion of the trawler masters hail from Ilhavo. The Portuguese merchant service draws many of its best recruits from this strange old town, whose foundation legend ascribes to the Greeks. Ships, ships, ships, sailors, sailors, sailors—Ilhavo's history is their history. In the 16th century, Aveiro—which the Ilhavo mariners considered a suburb of their town—sent more than sixty ships to the Banks, and had another hundred and fifty in general trade. Aveiro was one of the pioneer ports in the Newfoundland fishery.

Perhaps some of these things flashed through Captain Silvio's mind as he waited for Captain Leite to come to the microphone. He had an abiding interest in the maritime history of his country, especially in the history of its Transatlantic fisheries. The fisheries at Gloucester, in Massachusetts, were an offshoot from those of Ilhavo and the Algarve, in the days when Portuguese money was current in much of New England, and the Portuguese language was spoken in many places where it is now heard no longer, or only rarely. Those days are gone. Not many Portuguese now cross the North Atlantic to fish from Gloucester, as whole families did in his youth. Young Leite's grandfather was drowned in command of a Gloucester schooner on the Banks: his widow took the children back to Ilhavo and the boys were reared to become Bankers out of there. It might as easily have been Gloucester. Indeed, several of them went back to Gloucester, to fish from there, Leite's father among them, and English was spoken in the Leite home almost as well as their native Portuguese.

Young Captain Leite was typical of the modern Ilhavo Banks masters. Nowadays, instead of beginning as a deckboy and graduating as a doryman, the aspirant to command has a long period of more academic instruction, perhaps more theoretical than practical. But Captain Leite was still a Banks master in the old tradition, though he was not then twenty-five. He could handle the *Gazela* and get good speed out of her, though the only other square-rigged ship he had ever served in was the State's school-ship *Sagres,* a big barque in which he was but one of 300 boys.

"*Elisabeth! Elisabeth! Elisabeth!* Here is *Gazela!* Good morning, and good fishing, Captain Silvio." It was the voice of young Captain Leite. He was not much worried about the weather at the moment. His fishing, he said, was bad, very bad. (They all said that.) The dorymen were well away, and there was no present sign of any sudden coming of a southerly blow. But Captain Leite was worried about what he considered his poor catch of fish, and worried, too, about the salting. Were the fish salted sufficiently, or too much? Or perhaps not enough? There was no old-timer in the barquentine's after-guard to whom he could turn for advice. A first voyage in command was a worrying experience. The optimistic note that rang in his pleasant voice across the air in part belied the real anxieties he felt. He was, in fact, worried about the weather. The previous day had been full of mirage. The Greenland coast about the mouth of the Isortok fjord was contorted wildly as if it were trying to curl up and hide, and a great tabular iceberg inshore was miraged until it looked like a five-storied castle, sinister and enormous. He did not like mirage. It was a bad sign. He did not like calm, for that was a bad sign too. What good signs were there for a harassed fishing master in Davis Straits, with thirty-one dorymen's lives in his keeping? And a six-month voyage to make, such as his father, and his grandfather, and his great-grandfather had made before him. There was strength in that thought. What they had done successfully, he could do though ships had grown considerably and difficulties had not lessened with the passage of the years. When the older dorymen first saw the *Gazela* brought into service, rebuilt at Setubal in 1900 after seventeen years as a merchant ship, they were filled with gloom. This had been before Captain Leite's time, but the dorymen still remembered the presages of disaster.

"Too big!" they had said. "Too big, and too high! Dorymen will never fill that ship with fish!" Well, they had been doing it season after season, over the intervening fifty years. José Leite stared round the little white saloon, spotless and neat, with the old-fashioned compass swung in gimbals above the captain's place, *his* place, at table, so that it could be viewed from below and the captain never unmindful of the course; and the gracefully proportioned winding companionway which led to the deck. Through this there came more than a slight reminder of the Benz diesel with which the old barquentine had been fitted for the past decade. When her stern had been opened for the shaft, the timbers were as sound and as sweet as the day they were first put there. Four little cabins opened off the saloon, for the mate (aged twenty, and making his first voyage to the Banks and in a sailing-ship), the boatswain—then away in his dory—and the engineer, whose hands were full enough with the Benz, the refrigerator, and a diesel for the windlass.

Through the small skylight, Captain Leite could see the trysail set as a riding sail above the main boom. It was flapping now, for there was no wind. For how long might there be no wind? Captain Silvio had sounded anxious. He did not get anxious easily, though he was always concerned for his men. Captain Leite took his binoculars and hurried on deck. There was still no sign of any real change coming in the weather. All the dorymen he could see were fishing steadily, most of them hauling in their long-lines after the first cast of the day. He noted with regret that they did not seem to be taking many fish. Here and there, already he could see the tell-tale triangles of coloured sails breaking on the morning air, speaking of dorymen dissatisfied with the results of their first cast and setting off in quest of better ground. But the dories from the *Argus* were doing well. He could see a couple of them already returning to the four-master, full. Full dories were always for other ships!

Captain Adolfo Simoes Paião, Jr., of the schooner *Argus*, was thinking the same thing. His binoculars showed him four dories making for the schooner *Condestavel* and three more for the motor-ship *Cova da Iria*, while his own most expert men had broken out their sails and were heading not for the *Argus* to gaff up a fill of fish, but for fresh ground. Captain Adolfo knew all his dorymen even at a distance of three miles and more. All dories were alike, but their rigs were not. He could distinguish minor differences in the sails: the way Senhora de Oliveira sewed a mains'l was quite different from that of any of the other Fuzeta wives, and the Azoreans' sails were distinctive enough for anyone. If the sails were not set, Captain Adolfo was still able to distinguish his dories by the men in them—all fifty-three of them. His eyes did not linger on the two dories approaching the ship. One was the Little King's, the other belonged to a doryman almost as noteworthy for his poor catches. They were coming back, doubtless, because they had lost gear—grapnels, lines, hooks, bait. Always lost gear! One of the curses of the long-line fishing was that, when gear was lost, it was a costly loss. When a dory overturned, the financial loss might be considerable even if the doryman survived. In the old days, a doryman had his jiggers and his hand-lines, his personal pail, and the bait of his own catching, and that was all. But he did not go to Greenland then.

Captain Adolfo walked the confines of his short quarterdeck, dodging to avoid the skylights, the standard and the steering compasses, Setter, the bad-tempered water-dog, and Bobby, the yellow mongrel which was for ever round his feet, to say nothing of a couple of spare dories and a large salt-bin which the first and second mates were building on the starboard side to house salt from the hold and make room for more fish down there. More fish! More fish was what he wanted, what they all wanted. More fish, more fish, more fish! And here came the Little King, prince of bait-wasters, obviously to moan about lost grapnel and strong tides, and half his long-line lying off Holsteinsborg Bay.

During long years Captain Adolfo had acquired a good stock of patience. He had been in sailing-ships since he was eight years old, for his father had taken him as cabin-boy in the big wooden full-rigged ship *America*, at that age, and he had been a deckboy, not with his father, at nine. A cabin-boy from Ilhavo, with his father or not, had a hard life. Not only did he keep all the cabins in the poop spotlessly clean and tend the meals, but he had to help on deck too, and steer all day when the ship was in the trade winds and his elders at work. The *America* had a small crew. She leaked. She was old, and bits of her tophamper came adrift from time to time. She was in the trade between Oporto and the Gulf of Mexico; they were generally about two months from New Orleans back to Portugal. Before he was nine years old, he could take in a royal by himself in any but really bad weather, and he could patch old sails and splice small wire. Adolfo was one of five brothers who all became ship-masters, most of them in sail; but they scarcely knew one another, except when they were on the Banks together in their schooners. They had begun at sea at different ages, some with their father, others not. Once they went to sea, the separation was complete, though they continued to have homes at Ilhavo, and the Bankers among them were usually able to spend a few months each year there. The oldest brother, Manuel, had gone in the merchant service and had command of a deepsea steamer, and so was rarely home. Another brother had command of a Banks trawler which was at sea ten months of the year. The youngest brother, Julio, had died while in command of the Banks schooner *Cruz de Malta* a year or two before, leaving only Adolfo and his brother Francisco (called Almeida) of the *Creoula*, still with the sailing Bankers. Neither Adolfo nor Francisco had spent a summer at home, or any part of a summer at home for more than thirty years. They saw one another occasionally, at the blessing service, at St. John's and at North Sydney, for a few weeks before Christmas at Ilhavo. For the rest, their converse was by the radio telephone.

Well, that was the sailor's life. The seafaring man from Ilhavo was still a man cut off from shore pursuits, as the traditional seaman had always been but now was trying hard not to be, in so many merchant services. In Ilhavo he accepted the fact. The presence in the little town of so many deepsea ship-masters, officers, and men, and the acceptance by their womenfolk of their lot, was in many ways a source of strength to the Bankers, for they and their womenfolk were always among their own kind. Men had always gone to sea from Ilhavo: no girl really hoped to marry a stay-at-home. If she saw her man for three months of the year she was doing well. Even when they were at home, most of the men tended to keep to each other's company, talking about ships or looking at ships, down at Gafanha. A woman's business was to rear more sailors and properly to sustain those already at sea, though she was no longer expected to send her children away in ships at the age of eight. There were no more full-rigged ships and big wooden barques for them to go in, and the Gremio, among its many activities, kept a watchful eye on juvenile recruitment for the Bankers.

It was a long haul from the ship *America* and the barque *Clara* in the Rio trade to command of the steel schooner *Argus* in Davis Straits, and from the ancient two-masters of the old fishing fleet to the stately and powerful big schooners and motor-ships which came with the renascence of the industry. Captain Adolfo's lifetime spanned both phases, though he was a man only just past fifty. When he was a boy in the deepsea square-riggers, his father and his uncle did not go to the Banks. Banks fishing from Portugal was in a temporary decline and, moreover, the First World War had provided much employment for the mariners of Ilhavo, both under their own flag and that of Brazil. Many of them held licenses as Brazilian master mariners, as well as Portuguese. It was not until 1919 that Captain Adolfo first made a Banks voyage. He was 22

THE CAPTAINS FROM ILHAVO

years old when he first signed on as mate of the fishing schooner *Vencedor* but he had been at sea continuously for fourteen years. The *Vencedor* did poorly, taking less than one-and-a-half thousand quintals of cod; the life was incredibly hard, and the rewards poor. But the impetus given to the industry by the Government in the early 1930's and its reorganization under the Gremio, made a great difference. The men of Ilhavo flocked back into the Bankers, and they have been there ever since. Captain Adolfo, who in the interval had spent a few years in merchant schooners, was standing by the beautiful new *Hortense* on the stocks at Gafanha by 1930, to go out with her on her maiden voyage as mate. The *Hortense* was then the pride of the Bensaude fleet, and her first master was the famous Anibal Ramalheira. Anibal Ramalheira knew a born seaman and a natural codhunter when he met one: Adolfo Paião was both these things. From the *Hortense* he never looked back. His next ship was the shapely *Gamo*—another of the older Bensaude fleet, now gone —as mate; then the swift *Neptuno* in command. After her, *Hortense* in command, then *Creoula* and finally *Argus*, the commodore's ship. He had not lost a ship and, by the mercy of God, very few dorymen. It had been his good fortune, since joining the Bensaude line, to serve always in vessels of distinction: he had served them well.

But it was a strain on a man, a constant, nagging strain. Six solid months cooped up in a steel schooner— or a wooden one—year after year, tethered by the nose on the shallow, dangerous Banks, always with the lives of half a hundred men in his hands, required to fill the ship: his the decision which committed the dorymen, day after day; his the business of out-guessing the weather, of never being caught by the murderous sudden gusts of wild south wind . . . it *was* a strain. But it was his life. It was a man's life. Though they talked a lot about it, he did not think that many of the masters would change their fishing schooners for a more comfortable berth in the merchant service. Most had tried both. Almost any of them could go in merchant ships if they liked, and come Banking no more.

But year after year, the captains from Ilhavo sailed with the ships, bound out for the Banks. Year after year, anxious day after anxious day, they walked the small poops of their crowded schooners or strode the bridges of their motor-ships, worried about the lives of their dorymen, as he was doing now. His brother Francisco was doing the same thing in the *Creoula;* and so were Silvio Ramalheira in the *Elisabeth,* Armando Ramalheira in the *San Jacinto,* Manuel Ramalheira in the *Infante de Sagres,* and all the rest of them. All from Ilhavo. Men cut off from the land and its pastimes, on the Banks their only interest is cod. To them the world is the Arctic and the Grand Banks fishing grounds, and the very word "fish" means cod. The only ships are fishing ships, the only news of interest concerns these ships and the cod they hunt. Ilhavo is a place far away, from which they set out upon their fishing voyages. They rarely listen to world news for the radio is eternally tuned in to the discourse of their brother captains, and this is invariably of cod—of cod and bait and sudden blows, and dorymen and illnesses aboard and lamentations about the weather and the paucity of cod.

Aye, men cut off. The captains from Ilhavo were

THE CAPTAINS FROM ILHAVO

severed from the land far more completely than any doryman was. The doryman, year after year, retained his intense interest in his home and family and the things of the shore. To him the campaign was a necessary interlude, for his bread and butter. But for the captains, as the years passed it tended more and more to become their lives, almost their whole lives. A doryman's heart was always ashore, and there was no doubt about that. But a captain's loyalties had to be divided. In the end, in a good many cases, the ship won—the ship, the codfish, and the sea. There were old captains with the fleet who had come back again after retiring, for they found they could no longer live ashore. The doryman went over the side, but the captain, to a great degree, carried the real burden.

And now those two dories came back to the *Argus*, the dory of the Little King and his consort, and to the astonishment of all aboard and not least of the Little King himself, they were full. Not a Laurencinha full, it is true: but full enough. The Little King gaffed up his fish with an air of great superiority, called for more bait and a drink of cold water, and was gone for a second load, while Azoreans with full dories began to approach the ship from all directions and not an Algarvian was in sight.

"Ah, miséria!" Captain Silvio moaned, watching the scene through binoculars. "The *Argus* on fish again, and only a miserable few dories coming back here!"

Captain Adolfo, for the moment, held his peace. So did the south wind, and that was more important than the strange freak of luck which had put the Little King and a few Azoreans for once on fish.

CHAPTER THIRTEEN

THE GREENLAND CAMPAIGN—JULY

> And sleeping now was Mortals whole delight.
> Th' illustrious Captain (who had all that space
> Been kept awake about the last day's fright)
> Gave then to his tyr'd eyes a little sleep.

THAT freak day when the Little King filled his dory brought the *Argus* 280 quintals, for the Little King filled twice, and so did fifteen other Azoreans. It just happened that they found fish quite close to the ship, and the Algarvians—those confirmed individualists—had sailed off for miles, as they always did. It took even such experts as Laurencinha and João de Oliveira all day to fill one dory. Once the dories had gone, there was no way to indicate to the dorymen where they might find fish. It was each man for himself. To hoist the recall and bring them all back was no use, for at any time the shoals of fish might wander. The length of time necessary for a doryman to haul in his lines and sail back to the ship—or row back, if he must—was considerable and might well be wasted. Generally, the majority of the Azoreans kept to one side of the ship

The First Fisher—Francisco Emilio Battista, Called Laurencinha

ão de Oliveira Was Second Fisher of the *Argus*

Once a Doryman, Now Captain of the Schooner *Santa Isabel*—Captain Alfredo Simões, from the Algarve

Captain Silvio Ramalheira of the *Elisabeth*

The Afterguard of the *Argus*—All Save One from Ilhavo. (Left to Right) Captain Adolfo; the Mate; César Mauricio; Second Mate; Assistant Engineer

She Was the Barquentine *Gazela*

A Big Old Fellow from the Arctic Depths

The Second Mate

The Boatswain and His Son Had Survived the Sinking of the Schooner *Gaspar*

and the Algarvians to the other, though this was not always so. Off Isortok, off Umanak, and in Holsteinsborg Bay, the dorymen often had their own particular fancies: they all had the fisherman's instinct for bottom conditions.

But there were days when none could fill a dory, though they fished twelve or fourteen hours and cast their long-lines three and even four times. In the first week in July, despite that one good day, the ship did not take 400 quintals. At that rate, she would be fortunate to see Lisbon by mid-October, and there was great pessimism. When the fishing was poor the dorymen worked even harder, and it was most discouraging to them to jig all day, and cast and haul their long-lines, for half a dory-full. Poor old Manuel de Sousa, when his dory was hoisted in about half past six that afternoon—it was nearly full—sat down in the cold on a tub of wet long-line, and went to sleep while his son stood to wind'ard beside him, making a bit of a lee against the freezing wind. But he was splitting fish until past ten o'clock.

The *Argus* fished on Store Hellefiske Bank throughout July but her luck was rarely good. There was much bad weather, too much. We were obviously in for a bad summer. The old-timers said that in Davis Straits the weather in the summer months was usually either very good or very bad. It was seldom mixed. They had all known seasons when the dorymen had been able to fish without pause for six, seven, or even eight weeks without a break, launching every day and always finding fish, with the sea on the Banks like a millpond. They had known other seasons when they rarely fished a week without losing at least one day. 1950, it seemed, was destined to belong to the latter category. In the second week, the dories were aboard four days, and none of the other three was a full fishing day. The sea ran high and majestic and the wind howled in the wet rigging where the trys'l strained, and the ship had to be steered at her anchor with the diesel going, while she jumped and leaped and lurched. Captain Adolfo, anxious lest the storm should break loose icebergs grounded on the bank and send them crashing down on him, spent a worried four days. There was no peace for him, no peace for any of the captains. The dorymen, after making a thorough overhaul of their gear and checking all their hooks and snoods, slept in the rancho and, after the third day, even the deckboys could sleep, for there was no more work that they could do. But for the captain there was no rest. The visibility was poor, with driving murk; there were many other ships about, which could drag into him; the radio spoke of drifting ships and driving icebergs, and told of more than one narrow escape from destruction.

The wind was all very well while it blew directly up the straits, or down; but what if it set onshore? The rocky bottom of Store Hellefiske Bank was poor holding ground, and all Greenland would be a dangerous lee shore. Often, before a gale blew out, it did set onshore. Then the ships hung on. Already several of the older schooners were thrashing about under storm sail, their ground tackle carried away and their masters not disposed to risk the loss of more. We saw one of these schooners, a lovely three-master, clouding the sprays right over her and rolling until half her copper bottom showed, but riding the seas beautifully and shipping no dangerous weight of water.

"She is all right," Captain Adolfo said. "She's rolling her fish down."

Rolling her fish down? The mate explained that a certain amount of bad weather was desirable, for otherwise the salted cod did not settle so well and the schooners could not stow big cargoes. When the ship rolled and pitched she shook all the pickle from the cod and settled the fish down, as if they were being passed violently through a giant sieve. So the captains did not like too much calm. The *Argus,* for example, could stow varying amounts of salt cod depending upon the extent to which her cargo was shaken down. With a good shaking while she was catching, she could carry 14,000 quintals, though her official capacity was put down at almost 3,000 less than that.

We certainly shook our cargo down sufficiently in July and the dorymen with it, to say nothing of the deck fittings and everything moveable about the ship. It was extremely difficult to eat, even cod fillets and fried cods' hearts, though the imperturbable cook performed miracles. How that man and his assistants could so consistently turn out hot meals for seventy-one persons, day after day and week after week, no matter what the weather, was always a source of astonishment. He did not rule his galley with a rod of iron, as was the custom in sailing-ships, forbidding entry to all who had no business there. The galley contained the only fresh-water pump, and it was continually subject to a traffic of large dorymen who shuffled about and consumed draughts of cold water. The boatswain made his home there on bad-weather days, and Laurencinha had a favourite corner where he tested hooks. The single-fire stove served also to bake bread and rolls, and the amount of baking was prodigious—seventy small loaves at least, every morning, and plenty of excellent rolls. The baker on duty for the day began work at 1 A.M. and the other cooks slept until 3:30: two of them were generally still at work when the time came to serve the midnight soup of sorrow. The galley was not large. In addition to the stove, which was large, it contained coal-bunkers, a huge barrel of olive oil which was always kept ready for use, and an array of enormous pots and pans which hung from the deckhead and were stowed on racks wherever space could be found for them. The top of the stove was always occupied by one enormous cauldron which would hold seventy helpings of nourishing soup, and another with at least a hundred helpings of body-building stew, to say nothing of a huge coffee-pot and one of the world's largest tea kettles.

Despite all this, the cook managed, without fail, to turn out special dishes every day. His boiled fish and baked whole cod and fried halibut were always a delight. His inseparable companion was Amalia the fowl, who did not take kindly to being shaken down. In the bitter weather Amalia never left the galley, but remained perched as close to the large fire as she could get, preferably at the doors of the ash-pit. When she was not there she was on the shoulder of her friend the chief cook. Amalia was an undersized fowl with golden plumage and a trick of holding her small head in the position in which the cook had last stroked it, whatever that might be. She reserved her affections for the chief cook, Senhor Manuel Gordo Cardozo, and he always did his best to look after her. She survived the voyage but was not minded to make many more.

THE GREENLAND CAMPAIGN—JULY

At last the bad weather left us for a time, and the dories could be launched again, sometimes for half a day, sometimes for only a few hours. Some captains did not bother to launch much after midday and, if the dories were still nested then, they were kept aboard for the day. Captain Adolfo got his dories away whenever he could, so long as he could count on three hours of fishing weather. So did the other Bensaude captains and Captain Silvio. When the dories were launched late, however, the issue of bait was reduced, for it must be conserved. The dorymen were then catching fair numbers of the repulsive big catfish, with its leopard skin and evil eyes. Catfish flesh was good bait, and so was the flesh of halibut. Halibut could not be preserved satisfactorily by the usual salting methods and the refrigerator space was reserved for bait, so they generally found their way to the galley—halibuts' heads made excellent soup—or were used for bait. Each doryman kept his own. Most men kept at least two or three large catfish and several large halibut, each fishing day. The catfish were flabby, horrible fish, most repulsive to look at, but the men said their flesh was good. We did not eat it. Cod steaks, cod fillets, cods' hearts, cods' tongues, the membranous stuff from big cods' backbones, cods' cheeks fried in olive oil, whole cod, dried cod boiled with potatoes, shredded, made into patties, or fishcakes—these kept us going, with now and again some fried flounder or halibut. No one tired of well cooked fresh fish.

Day followed day and our cargo mounted slowly, far too slowly. Few cargoes of cod in salt bulk could have had a better shaking, and the fish hold seemed to grow more cavernous and empty every day. The deckboys still found more salt to shift, moving every single crystal. The two mates were the carpenters, and they built up temporary wooden salt bins wherever it was possible to find space for one—atop the for'ard house, where it further restricted the helmsman's already poor field of view and cluttered the foresail so much that the sail could be set only with the greatest difficulty; on both sides of the quarter deck; in the only clear space on the main deck, immediately abaft the mizzen mast. Even these extra stowages were not enough for the never-ending stream of salt pitched up from the compartments in the hold. The two big dories on the skids for'ard were cleaned out and filled with salt. So was the port coal bunker; and in due course, as the bait stocks went steadily down, the starboard refrigerated room became yet another salt stowage. The aim was to keep two compartments in the hold always ready for the reception of fish. As one compartment was filled with cod stowed on their backs, fore and aft, it was left for a time until the fish were shaken down. Then it was restowed full once more, shaken down again, restowed again, by salters lying on their backs on a large tarpaulin. They carefully placed each fish in position and made sure that each had its share of salt and no more.

There seemed no end to the capacity of the *Argus* to stow fish, provided she rolled enough. Long after the lower hold should have been filled, compartments which first took fish on Fyllas were still being restowed. This was a skilled and laborious business, and sometimes special men were kept back to get on with it, even on days when the dories could be launched. The deckboys only manhandled the salt, shifting it from one place to another. They did not touch fish. They had work enough without that: the thorough cleaning which the fish-deck got every morning took them at least four hours. Pounds, vats, chutes, trolleys, gaffs, splitting tables, the deck itself—everything was scrupulously scrubbed clean. They were rarely able to begin this work before seven, though they were called at four. Getting up bait, helping to launch the dories, cleaning cods' backbones, putting tongues in brine, salting cods' cheeks—these were their first jobs. A Banker's deckboy led something of a dog's life, though they were cheerful boys and young men.

One at least of them knew what it was to swim for his life from a sinking schooner. The boatswain's son had been with his father when the *Gaspar* was lost, and they had both survived by a narrow margin. The *Gaspar* was old and wooden, and had been assailed by much bad weather. The boatswain's son was very thankful to be in a steel vessel. Another deckboy was the son of an Azorean harpooner, a man who still hunted sperm whales by hand from an open boat. I asked José Cabral whether he would not prefer to be with his father in this stirring business. But he grinned and said "no": there would be time enough to harpoon whales and go to the Banks as well, for the whaling was a most irregular business. He had been with his father many a time, but it was his ambition to become a first-class doryman, and he had attended the professional school for young fishermen at Lisbon with that in mind.

The professional school was yet another institution of the industrious Gremio: every ship on the Banks had some boys from it, and an excellent lot of young fellows they were. They worked indefatigably and were always cheerful. Perhaps the hardest-working of them all, in the *Argus*, was the cabin-boy, who had his meals aft and in return for that privilege was called half an hour before the others and got to sleep at least half an hour later than most. He not only kept six cabins, the saloon, and two long alleyways spotlessly clean, but was in general charge of the tools, helped to tend dories and to work salt and, by night when not otherwise employed, worked on the tongues. The cabin-boy was a young man from Ilhavo, aged perhaps twenty and newly married. Even on days of bad weather, he had not much chance to rest for there was always extra work in the saloon then. When they were not doing anything else, the deckboys helped the dorymen to clear their lines, which continued to become badly snarled on the rocky bottom.

All the deckboys would rather have been dorymen, despite the risks of that calling and the hardship of having to work so close to the sea all day, standing on the frail planks of a small flat-bottomed boat. It was better than shovelling salt. The pay was considerably better, too. A doryman had what he earned; a deckboy received his wage according to a fixed scale which was not high.

One day early in July, there was yet another instance of the simple manner in which a doryman might lose his life. The Azorean doryman Baleia, the Whale, was spilled into the water through a slight error of judgment which came near costing him his life. The weather was not good and not bad. The day was dull with a sighing wind and a white, greasy breaking on the tops of the seas, fishable for good dorymen, but not at all pleasant. When the dorymen began to come back, Captain Adolfo lifted the anchor and let the ship drift broadside to the sea, so that the starboard side was a good lee and the ship acted as a breakwater. The Whale, a comparatively young man without much experience, failed to observe what had been done. Coming back with a good dory-load, he tried to cross the bow and make for the Azorean side, which was to wind'ard, with the sea breaking along it quite dangerously. He did not reach the Azoreans' side, for he misjudged his turn at the bow. His dory touched—just barely touched. Over it went, and the Whale was spilled into the sea with his white-bellied cod floating all round him. Captain Adolfo had shouted a warning to him not to go to wind'ard as he passed the skids, but the young doryman was wearing heavy ear-muffs and could not hear.

When his dory overturned, the Whale came to the surface, spluttering, and struck out at once. He was right alongside the ship, under the flare of the bow, but that was as good a place to drown as any, and he was very heavily clad. Fortunately for him another doryman was close by with a light load of fish. This doryman hauled the Whale aboard, over his stern; other dorymen began immediately to salvage the gear which was floating away from the overturned dory, and to save the dory itself. As for the Whale, he was at once brought aboard, to receive a dressing-down for being such a fool as to try to make the dangerous weather side when he had reached the lee and been told to stay there. For a moment he looked almost sorry to have survived, but a change of clothing, a rub with brandy, and a pot of coffee soon put things to rights.

He was on deck again within a quarter of an hour, looking to the re-assembly of his dory's gear, which had all been salved by then. Everyone at hand had helped to salvage things as they drifted past, the cook taking a leading part in these activities. All three cooks were always on hand for emergencies when the dories were coming back. Senhor Gordo Cardozo, grasping a long-handled boat-hook, first saved the Whale's personal pail, then his bait-basket, long-line tub, and sail. Old Jacinto Martins salved the long-line itself, by means of its buoys, for the line was adrift in the water. Other dorymen picked up the thwarts, oars, compass (afloat in its wooden container), bailer, and the anchor-line of the dory. By this it was hauled back to the ship, righted, bailed out, and soon hoisted inboard. Some of the older men, when they feared the broken water by the ship, took the precaution of handing over their heavier gear to friends less heavily laden and, sometimes, even part of their fish.

When he was thrown into the water, the Whale was dressed in check flannel underwear, heavy woollen trousers—also of check pattern—two pairs of woollen socks, flannel seaboot slippers, two check shirts, a jersey, and a second check shirt outside that, rubber thigh sea-boots, an oilskin suit, a balaclava, ear-muffs, and a sou'wester. He lost none of this gear. Indeed, he was into the water and out again so quickly that some of it was scarcely wet, for his oilskin suit was good and secured with soul-and-body lashings. A man dressed like that, however, needed to get out of the water quickly if he was to live. All the dorymen dressed more or less like that on the Store Hellefiske Bank.

There were some good days in July as well as many

bad, days when the calm lasted and brought neither fog nor sudden increase of wind, and the fishing was good, and even the captains were not so pessimistic as usual at their nightly conference over the radio telephone. There were days when it was so quiet that dogs could be heard barking aboard schooners miles away, and Laurencinha could bring back his dory with no freeboard at all. Gradually the whole of the handlining fleet assembled on Store Hellefiske, with the last St. Malo-man also in company. The French Banker kept herself to herself and fished her own way, with her two-men dories and her systematic manner of laying out the long-lines. She was never nearer than two miles away. Through the binoculars, the unmistakably French lines of her steel hull could clearly be seen: she was a saucy looking ship, with a long bowsprit, steeved high. She had been a barquentine but now there were no yards on her foremast, and very few sails on any of her masts. The *Lt. René Guillon,* according to all accounts, was fishing very well, but she had been on the Grand Banks, before coming up to Davis Straits, longer than most of the Portuguese vessels, and—rumour said —had taken at least 5,000 quintals there. She had a long poop to accommodate her people. She had not to carry nearly as many dories as the Portuguese schooners of comparable size. Her mizzen mast was an engine exhaust, and we could see a large loop-aerial for a radio-direction-finder on top of the charthouse on her poop. She had a fine clipper bow, but otherwise there was little of the sailing-ship about her. She was operating practically as a motor-ship, and the few sails she had bent could be riding canvas only. Most of the gaffs and booms were gone, as well as the yards.

THE QUEST OF THE SCHOONER ARGUS

We wished her luck: there were fish enough for all the hand-liners, and still the trawlers stayed away. They were doing well on the Grand Banks, we heard, and so were the schooners still down there; so well, indeed, that a few of the smaller Greenlanders wished they had thought of acting on the old adage "Bad ice, good cod," for the good cod were there, right enough, and they were not abundant in Greenland waters.

But no one shifted back to the Grand Banks—at any rate, not then. It was too far to go, and there were fish enough on the banks in Davis Straits. The trouble was the weather: a schooner might easily waste the only two good weeks of the summer trying to sail back to the Grand Banks, and arrive there in time for the hurricane season and little else. Captain Adolfo did not consider the idea. We would stay where we were and fill on Store Bank, or not fill at all. If we had to try the Grand Banks again, it would not be until we had first been driven from Greenland.

All this was conducive to gloom. The *Condestavel* was reported to be doing well, and so were the *Elisabeth* and the pretty *Senhora da Saude*, a four-master which had been built in Denmark. None of their masters ever spoke of going back to the Grand Banks. There were several masters who rarely spoke at all, and then only when spoken to—Captain Chuva o Anjo (the Angel) in the *Hortense*, Tude Namorado in the *Senhora da Saude* and a few more. But the silence of these was more than offset by the volubility of certain others. There was one motor-ship captain who rarely left his microphone all day, and must have waked in the mornings with the thing alongside his pillow. It was his one diversion. According to him, he never found any fish;

THE GREENLAND CAMPAIGN—JULY

all his dorymen were hopelessly incompetent, and his motor-ship was a compound of melted-down airplane parts which worked alarmingly in anything of a sea. This pessimism was all part of the trade, and many of the masters seemed to compete for the proud distinction of being the greatest spreader of despondency. The dorymen had nothing to say but, among themselves, fishing on the banks, they picked up more reliable news than ever came over the air. After all, every doryman had a very good idea of the capacity of his ship, and he knew how she was doing. They often met dorymen from other ships and news got round.

By mid-July, we had fifteen ships in company, all, with the exception of the St. Malo-man, Portuguese hand-liners. The group was anchored well out on the bank, off Isortok, which according to all previous experience, was a good place for cod. But the cod continued to be playful. Sometimes they were there and sometimes not. A schooner a few miles away might find fish plentiful, and others near-by take not one full dory-load all day. The fish were most uneven. Some were enormous, others very small. All were ravenously hungry but, for some extraordinary reason, sometimes they would pass the bait.

One very quiet day, an Eskimo boat came out to us, looking for bread, tobacco, paint, fishing gear, and anything else that was going. The Eskimos had been fishing with a fifty-hook line, and we saw that their bait was lance, half a lance to each hook. But they had less than twenty cod in their boat, so they were not doing very well, either. There were three of them in a remarkably decrepit motor-boat perhaps sixteen feet long, the engine of which had to be warmed with par-

affin flames for a long time before it would run. The
Eskimos wore sealskin trousers shaped like jodhpurs,
and they all had hooded jackets though the day was
quite warm. They spoke no language but their own,
not even a word of Danish, but their borrowing expedition was successful, despite that. Their grapnels were
stones and their boat looked unfit to venture outside a
sheltered fjord. Later we were to see more enterprising and better equipped Eskimos, but this first party
made a poor impression. They had a couple of broken-
down shot-guns in their boat and had been shooting
seals. Captain Adolfo told them to bring out some
salmon from the fjords, but they pretended not to understand. There were plenty of salmon inshore which
would have made a welcome change from cod.

The second part of July offered some better weather
but the fishing showed little improvement. By the end
of the month, the *Argus* had less than half a cargo
stowed down below, with only another month of tolerable weather to hope for, if that, in Davis Straits. The
fish seemed harder to catch than ever.

"They are all lance-crazy," Captain Adolfo said.
"They won't look at our bait of herring or mackerel.
And we've nothing else."

He meant that the lance were running on the banks,
inward bound for their spawning grounds, and the cod
knew it. The cod were hunting the lance voraciously,
often coming to the surface to scoop up great mouthfuls of them. We caught cod which were so full of
lance that, when they were landed in the dory, live
lance leapt from their throats.

If the cod left the bottom to hunt the lance, there
was little use trying to catch them with bottom gear,

and the long-line was so contrived that it could only be made to lie along the bottom. So, for a few days, all the tubs of long-line were taken from the dories, and the dorymen fished all day without bait, with their jig-lines. This was immensely hard work, but it had to be done. This method of fishing was quite different from that previously used. The dories were launched together and stayed together, the ship weighing with them and drifting along in the current, to keep them company. Sometimes the ship remained at anchor part of the day, while the dories worked over a particularly productive area. They would drift along, the doryman upright in the boat, jigging furiously with both hands. Then, when he had drifted a few miles past the ship, he would haul his lines in and row back up-current again, to repeat the performance. As the current was running three or four knots, rowing was hard labour.

It was harder to fill a dory with the jig-lines than with the long-lines, for every cod had to be caught individually and hauled up from thirty fathoms down. If the cod were scarce—as they frequently were—it was a most discouraging business. But the dorymen kept on doggedly, pulling their arms out. *Paciéncia!* Sooner or later, patience would be rewarded. But it took all day for even the most expert to fill his dory: some days, no one could jig a fair load. But on they went, hour after hour, indefatigably. Sometimes schools of lance came near the ship, swimming very near the surface, but we never had the good fortune which some ships had known a few years before, coming up from Fyllas, when the sea was full of lance and so full of cod after them that the cod could be hauled into the dories by the bucket-full, and many of them literally jumped into the dories. No, we had no luck like that. That kind of thing always happened to other ships, or on other campaigns. I did not see any cod leaping into dories.

"And you won't," said the mate. He knew what he was talking about.

CHAPTER FOURTEEN

CRUISE IN THE GIL EANES

> And one lookt weather-beaten and halfe-drowned,
> As if a longer voyage Hee had gone . . .

BEFORE the middle of July, the *Gil Eanes* had arrived on the Greenland grounds to give assistance to the ships there, and to care for their sick. She had sailed from Lisbon early in May and gone first to the Grand Banks, to look after the ships there. Then she had discharged a cargo of salt at St. John's, Newfoundland, gone down to the mines at North Sydney to take in enough bunker coal for two and a half or three months, and come up to Greenland. The *Gil Eanes* was an old cargo ship of 1840 tons gross, 1048 net, and she could carry about 2000 tons of cargo, even with half the for'ard 'tweendecks converted to a forty-bed hospital. Financing an assistance and hospital ship is expensive, and so she always carried a cargo of salt from Portugal to Newfoundland at the beginning of her assistance cruise, and took back a cargo of Newfoundland salt cod at the end of it. In the winter months she continued to make this haul—salt out, cod back—in order to earn as much as possible while the schooners were laid up.

Though a small ship for the wild North Atlantic, the *Gil Eanes* was a remarkably sturdy and able vessel. She had to be, to keep the open sea for six months of the year and to battle to the westwards over the Atlantic in winter, deeply laden with salt. She was built by G. Seebeck at Bremerhaven-Geestemunde early in 1914 as the steamer *Lahneck* for the German Hansa Line and, during the first World War, had been taken a prize by Portugal in the Azores Islands. She was first a naval transport and she still belonged officially to the Portuguese Navy. She was managed now by the Gremio's owning organisation, called the S.N.A.B.—the *Sociedade Nacional dos Armadores de Bacalhau*, the National Society of Codfishing Shipowners—which, in addition to the *Gil Eanes*, managed the schooner *Oliveirense* and six of the largest of the Portuguese Banks trawlers, all named after pioneers of Portuguese transatlantic voyaging—*João Corte Real, Alvaro Martins Homem, João Alvares Fagundes, Pedro de Barcelos, Fernandes Labrador, Estavão Gomes*. Gil Eanes himself, for whom the hospital-ship was named, was one of Henry the Navigator's captains, the first to pass Cape Bojador on the west coast of Africa when the Portuguese pioneers were seeking to find the way to India by the Cape.

Something of the brave spirit of Gil Eanes still lingered in the old ship, which had survived more than her share of Grand Banks hurricanes and North Atlantic storms. She had been making assistance cruises since 1927. Hospital ships, of course, were no new thing on the Grand Banks, for both the British and the French had formerly commissioned such vessels to look after their fishermen. Some beautiful little schooners had flown the French flag on assistance missions. But the *Gil Eanes* was the last real assistance-ship to make regular voyages to the Banks, and the only such vessel which went to Greenland. She was, moreover, the floating centre of an organisation for the care of the Banks fishermen such as no other nation had ever conceived. The three thousand men in the sixty-four Portuguese fishing vessels then on the Banks and in Davis Straits were all medically examined before leaving Portugal; there were two surgeons and twelve male nurses scattered through the fleet to look after them; and there were two surgeons and two nurses ready to accept serious cases in the *Gil Eanes*, with which all the ships were in constant touch by radio telephone. The medical and spiritual care of the men did not begin and end with the *Gil Eanes*, which was only part of a national organisation caring for the men and their families. She was the floating hospital, the carrier of spares and of stores, the bringer of mail and of parcels from home, the wireless link with the world. Furthermore, she was the abiding and substantial evidence of the very real interest of the whole people of Portugal in their Banks fleet and the men who manned it. All Portugal indirectly contributed to her upkeep, through the payment of taxes on dried cod. She was manned by seventy men, at the head of whom—and in general charge, as the Government's representative, of the whole fishing activity—was Commander Americo Angelo Tavares de Almeida, a former captain of submarines in the Portuguese Navy, in which he was still an active officer though he had been coming to the Banks for seven years. Commander Tavares de Almeida was Captain of the Port of Davis Straits and the Banks off Newfoundland, and administrator of the fishing laws. In command of the *Gil Eanes* was one of the Ilhavo Ramalheiras, Captain Vitorino, himself a veteran of the Banks, a qualified Brazilian and Portuguese master mariner, and a master codhunter in his time. He had commanded the *Creoula* and the *Argus* and the *Neptuno* and *Gamo* before them. He had been Mate at nineteen. He had lived in the United States for some years, and sailed out of there, and his knowledge of the English language was excellent. Commander Tavares de Almeida, too, spoke excellent English, and this was the language generally used on the radio telephone when speaking to ships of other nationalities, except French. Most of the educated Portuguese spoke excellent French.

Captain Vitorino, when in command of the *Creoula*, had once fished 640 quintals of cod in a day with fifty-two dorymen, and this stood as a record. The *Creoula* was fishing just south of Holsteinsborg at the time—it was in the early summer of 1939—and full dories began to come back to the ship before eight o'clock. Most of the dorymen filled their dories three and four times that day, and the decks were so full of fish that the cod spilled over the windlass and piled high to the booms. He had called a halt to the fishing at three in the afternoon and, though they began splitting then, it was eight next morning before all the fish were salted down. Then four hours sleep, and away again. The *Creoula* had taken 1200 quintals in three days. The *Argus* was in company and took a thousand quintals in the same time.

Captain Vitorino had been in command of the *Gil Eanes* since 1945 and no longer sent dorymen off at four in the bleak mornings. Now he dodged icebergs, found elusive schooners, performed astonishing feats of seamanship almost daily, taking up precarious berths, carrying diesel oil, fresh water, galley coal, fresh food to leaping schooners in the open sea, or taking aboard seriously ill men whose dories became, for the moment, their ambulances and their stretchers. Sometimes he took the ship into some grim and rocky fjord along the coast of Greenland, to provide a brief period of quiet while the surgeons operated on a stricken doryman; or piloted the *Gil Eanes* into one of the minute and intricate harbors along that coast—Sukkertoppen, Holsteinsborg, Godhavn—to replenish fresh water, land the seriously ill, even to bury dead. Sometimes he performed heroic feats of salvage, as when the schooner *Labrador* was partially dismasted, or when seven schooners were driving upon the savage lee shore of Newfoundland in the height of a hurricane which had gone wrong and re-curved towards the land. Command of an assistance ship was no sinecure.

Nor was the post of Admiral of the Davis Straits and Chief of Assistance for the Ships at Sea, which Commander Tavares de Almeida held. But the Commander, even when things were at their most hectic—as they often were—breathed a sigh of relief that at least there were no convoys now. The *Gil Eanes* had led the sailing convoys in the recent war, after the *Maria da Gloria* and the *De Laes* had been sunk. Controlling a sailing convoy of anything up to thirty schooners of widely varying capacity, tonnage, performance, sail area, and crew, must have been a nightmare. The Portuguese Bankers had formed the last sailing convoys. There were usually two each way every season; the older, slower schooners to the Grand Banks, and the bigger schooners, with the motor-ships, direct to Davis Straits. Each group went back to Portugal in its own convoy as it had sailed. Station-keeping was managed by sail-handling, particularly by means of the large fishermen's staysails set between the masts. Usually the sailing convoys kept to themselves, all the ships clearly marked with the Portuguese colours painted on their sides, bows, and counters; and all burning their navigation lights by night, unlike the great hurrying crowds of ordered steamers, deep-laden and silent, which they sometimes encountered. Commander Tavares de Almeida remembered one occasion when some of the schooners became mixed up with an east-bound steamer convoy of upwards of a hundred ships, with the wind blowing hard from the sou'west, and the visibility murky with driving rain. The agile schooners, lean and fleet, twisted in and out among the columns of ploughing steamships, while destroyers, leaner and fleeter, tore after them in case they should prove to be surfaced U-boats in disguise. Getting mixed with the unlighted convoys by night was even worse. Many officers in the escorting vessels seemed reluctant to believe that there could still be sailing convoys on the North Atlantic in the 1940's, in time of war. The white schooners were lit by many a starshell and examined anxiously, although they never sailed until all belligerents had been informed.

Now those days were over—at any rate, for the time

being. Commander Tavares de Almeida and everyone else in the fleet hoped devoutly that they were over for ever. But the radio was speaking of a war in Korea, a "small-scale" war at the moment; but it might flare up into something big.

At first, when she reached the Greenland grounds, the *Gil Eanes* hurried among the fleet taking off the seriously ill. When she spoke the *Argus,* I joined her for a week or two, to have a more general look at the fleet and the Greenland fishing activities. She seemed a big ship among the fishing schooners, with her lightly laden hull and shining white paint, her high bridge, and her black-topped yellow smoke-stack bearing the badge of the S.N.A.B. The weather was tolerable that day, and I joined by dory. Aboard, it was obvious that the old steamer was a ship of character. I could hear pigs grunting somewhere, and a cow mooed. From an open hatch on the after deck arose a great cackling of hens as if there were hundreds down there, as indeed there were, and several cows, sheep, and pigs with them. This was the fresh meat supply for the hospital, for the *Gil Eanes* had no refrigeration. No matter, a fresh-laid egg was acceptable in Davis Straits, and thick steaks helped to build up sick dorymen. A black-whiskered sailor acted as farm-hand in charge, and all the beasts looked fat and remarkably contented. There was plenty of hay for them. From the galley, tempting odours rose; and the slow pulse of the reliable old steam engine, panting powerfully on its bed down below, spoke of power to spare.

"But she burns much coal," the Chief Engineer said. "Aye, aye—much coal. She is a very old ship."

She was spotlessly clean everywhere, from the radar

THE QUEST OF THE SCHOONER ARGUS

tower on the bridge to the jackstaff at the bows and the log-bracket fitted on the rail aft. When she was stopped (which was fairly often, for she had to visit many ships), all hands got out jig-lines and began at once to jig for cod, which they cleaned and salted for themselves, to take home when the ship got back to Portugal in late October or November. The main-deck was littered with small boxes to stow salt fish, and there were more jiggers flailing from her rail than from any schooner. Cooks, trimmers, firemen, able seamen, nurses, doctors, mates, stewards—all were busy jigging, as purposefully as Laurencinha but not often with the same success. However they all caught some cod; and the gear strung up below the cargo derricks was festooned with fish, drying in the wind. If the *Gil Eanes* remained at anchor long, her scuppers filled with fish.

Meanwhile the old ship hurried round the fleet, first to the schooners *Adelia Maria* (a pretty four-master) and *Condestavel*, which had a doryman spitting blood. The dorymen were loth to declare themselves ill, and more loth to join the *Gil Eanes*. They knew they would be kept there until they were well, and there was no fooling the young doctors. A doryman hated to be idle, and he hated, too, to be laid-up just when he should be at his busiest, earning a living for his family. Among the many innovations the Gremio had introduced was sick pay, and a doryman no longer lost all his pay while he was ill. But he could fill no dories, and his gross of fish could not mount. Tuberculosis was the major scourge, and there were two hopeless cases in the small isolation ward. The worst was a doryman from the *Soto-Maior* who had been adrift in fog for thirty hours on the Grand Banks. Poor devil, he would go adrift no more.

As she came to each ship, it was customary for the captain to board the *Gil Eanes* to pay his respects to the Commander, and to report his take of fish to date, for the Gremio's records. This, many captains were most reluctant to do. To find out their take of fish was almost as difficult as to induce a doryman to admit that he was sick: at the best, they offered a pessimistic estimate which was often less than the real total by at least a thousand quintals. In vain was it pointed out to them that the figures were confidential, and required only for insurance purposes and for marketing forecasts. They did not care about insurance. What mattered was secrecy about the fish. That was the tradition, and they intended to stick to it. None boasted; none was ever optimistic, even when his ship was bulging with fish and he might well have prayed for a gale or two to shake his cargo down.

"Oh, wretchedly few fish, Commander!" was the general lamentation. "Too much bad weather, and the cod all lance-crazy! We shall have to go back to the Grand Banks when we have finished here, for we shall never fill."

I saw Captain Vitorino grin. But it was said that he had been one of the biggest pessimists in his time. They all complained about their miserable fishing until the last moment, when their ships were so full that not another cod could be jammed aboard. Then it was their delight suddenly to turn into amiable optimists and to proclaim loudly over the radio telephone that they were full and about to sail for Portugal. It was a game, and they all played it.

THE QUEST OF THE SCHOONER ARGUS

They were an interesting group of men, those Portuguese Arctic captains—men strong of countenance, lean, full-bearded, dark, handsome; some dressed like fur trappers, none in brass buttons, few even with a braided cap. A fur-lined cap from Newfoundland or a tarred sou'wester was their headgear, and their clothes were thick and serviceable, for they always came by dory and left the same way. Their wants were varied but generally included fresh vegetables, often a spare anchor and some cable (part of the after 'tweendecks where the beasts lived was a veritable anchor shop), or a generator put to rights. They often brought deckboys to have bad hands dressed or bad teeth extracted, which was done at once without ceremony. Deckboys were not reluctant to visit the *Gil Eanes,* as the dorymen were. They had less to lose, and a week or two in a hospital bunk, with fresh meat to eat and chickens now and again, was what they hoped the doctor would order. This he rarely did. The two young surgeons in the *Gil Eanes* were a capable, hard-working, and most conscientious pair, and they knew deckboys. They knew dorymen, too, for neither was making his first voyage.

The hospital-ship gave them a most varied experience. On the previous cruise, for example, they had had to treat nearly 800 patients and perform thirty-six operations, several of them serious. The operating theatre was a small steel box on the fore part of the 'tweendecks and, though there the pulse of the ship beat softly, there was plenty of motion. A major operation was performed only when it was really necessary, but that was often enough. They did not treat Portuguese fishermen only. The *Gil Eanes* was, and had been for years, the only hospital-ship regularly and actively employed on the fishing-grounds and in the North Atlantic. Her services had several times been urgently called for by large vessels and, earlier in the year, she had saved the life of a man in a big French steamer during an Atlantic blow. The French had a frigate cruising the Banks but she was not designed either as a hospital-ship or as an assistance-vessel, and the *Gil Eanes* made her services as freely available to the dorymen of the *Lt. René Guillon,* and the crews of the Spanish, Italian, French, and British trawlers, as they were to ships flying her own flag. Her assistance was international. No wonder the dorymen all waved to her cheerfully as she passed, or blew a welcome upon their conch shells. To them she towered like an ocean liner, and her bow-wave was a minor bore. To them she was counsellor and friend, though she often had to pick a way warily among them, especially in fog. Fog kept no dories nested and it did not keep the *Gil Eanes* at anchor, though her new radar could not be relied upon always to indicate such small objects upon the sea as a fourteen-foot dory. A sharp lookout was kept at all times. The radar had been installed only that year. It was invaluable for giving warning of ice, but the profile of a dory was lost in any breaking sea.

Often enough in fogs, there were appeals from schooner captains for help in the search for their lost dories, adrift by night in the strong tideway, and always the *Gil Eanes* joined in the search. The anxiety in the captains' voices as they sought news of their missing men belied all the light-hearted complaints which had been heard before. They were profoundly upset at the idea of losing men, and knew no rest until all were back or accounted for. It was not at all unusual for a score or more dorymen to be adrift all night. When that happened, the radio on the fishermen's wavelength cried until the men were found and, until then, there was no peace of mind aboard the *Gil Eanes* either. The dreadful anxiety in all the voices as they asked ship after ship for news was pitiful to hear. Two men were adrift from the *Argus,* two from *Elisabeth,* twelve from the little *Labrador.* Poor captains! Their souls were twisted and the agony wrung from them, on this job. And poor dorymen! The combination of sudden fog, fresh wind, rising sea, and strong tide could be too much for any doryman. Their compasses might become unreliable, so far north, not far from the magnetic pole; their dories were always overloaded, if they could possibly contrive it so; their ships were small and hard to find in the wide straits. The fog was dense, and the sea always rose quickly from a deceptive calm to a breaking turmoil most dangerous to loaded dories. Near the edge of the banks a dory might easily find itself driven into water where its forty-fathom line would not reach the bottom. The greatest hope was to anchor and ride it out; but off the banks you could not anchor.

That night the St. Malo-man found the *Labrador's* dorymen in company together, and by midnight we heard Captain Silvio announcing that he had the two from *Argus.* The *Lutador* had found eight of hers, and another two were in a trawler. The weather cleared a little after midnight, and then all the dorymen were accounted for. It might so easily have been otherwise.

CRUISE IN THE GIL EANES

I thought of the long lines of cold and tired men cleaning fish aboard the *Argus*, toiling mechanically far into the night, and trying hard to keep their eyes from the gaps in the nests where the missing dories should have been and their minds from thinking of the men who could so easily have joined the man from Nazaré, who had sailed off bravely in his dory one fine Greenland morning and never returned. It was late when Captain Silvio announced the rescue, but the news lifted a heavy weight from tired minds. This kind of strain was commonplace but the men never became accustomed to it. The very thought of a man adrift in fog and wind, when he might be lying-to in safety and sheltering under his oiled sail, was always unnerving in the savage cold. The cold and the weeping fog, and the never-ending, exultant moaning of the icy wind, could get upon their nerves.

These periods of agonising uncertainty were only too frequent: sometimes as many as thirty dories were missing. No matter how many times they had been lost in the past and found again, it was always possible that this time they would not be found. Now, with the radio telephone, the chances were better but they were still chances. The murderous sea could snatch a dory so easily, and the doryman's cries would soon be stilled. The dory would come to the surface again but the doryman would not, and he knew it.

One wild night the *Gil Eanes'* radar had guided her to the *Elisabeth*. Fog had blown down after a clear day, just as the dories were being called back. Then the weather deteriorated, steadily and rapidly, with a short, steep, breaking sea. The fog was dense, the wind icy. The *Creoula* was not far away. Many hours after they should all have returned, both she and *Elisabeth* still had dories adrift. At once the two captains appealed to the *Gil Eanes* to scan by radar, in search of the missing men, but it was impossible to detect the profile of a loaded dory when the seas were so high. Hour after hour, *Elisabeth* and *Creoula* cried on their great sirens and with their bells, while the *Gil Eanes*, continuing her radar search, roared on her steam siren. Sound plays strange tricks in fog: the dorymen might hear any of these sounds or they might hear none. Some of the *Elisabeth's* dories at last returned and, by 10 P.M., all the *Creoula's*. But three from the motorship were still adrift.

At last, towards midnight, the three little fellows were suddenly beside us, where the *Gil Eanes* was at anchor close upon the quarter of the unseen *Elisabeth*. They came together silently, at first three small sails, indistinct as the fog itself and seeming like wraiths adrift in it. But the ghostlike sails grew into real sails, and there were the dorymen. Two of them looked up and grinned; the third man hailed. "Where is *Elisabeth*?" "Close by alee" came the answer from a dozen throats, for practically all hands had remained about until the dorymen were found.

The dorymen were standing upright in their heavy-laden boats, two of which were rigged with dipping lugsails and the third with a small blue mainsail and a tiny jib, the red cross of Christ emblazoned on the main. In yellow oilskins and oilskin skirts, each doryman was holding his steering oar alee and the mainsheet in his hand, securing the sheet at times to throw out blood-red water, furiously. Dead cod filled the after parts of all three dories and lay between the thwarts for'ard. Tubs of long-line in the bows, little anchors neatly stowed flukes for'ard, empty dinner-pails atop the fish—everything was shipshape and seamanlike. The tide ran hard, for it was the time of springs. It was against the wind, and the nasty sea it caused to run upon the Banks broke constantly at the gunwales' edge of each small boat. Yet the dorymen looked up and grinned when they knew that they were safe. To me their daily survival seemed little short of a miracle, but it was a miracle which must be repeated upon the morrow—upon fifty morrows. There is no recorded case of a doryman losing his nerve. No man ever came to the *Gil Eanes* with any kind of nervous complaint. Their occupational diseases were pulmonary, and of the stomach.

Those dorymen had been making an organised search for the *Elisabeth* for more than four hours, taking care always to keep to wind'ard. Each time they ran their estimate of the distance to the place where their ship should be, they anchored for a few moments, to test the current; then sailed back upon their tracks, or as nearly as they could, to begin again, and run the estimated distance once more. Sometimes they could hear the siren of a schooner or a motor-ship, but for hours they could not find any ship. The visibility was less than a hundred feet. In the end, it was the bull-roaring of the *Gil Eanes'* steam siren which saved them. They homed on that once, twice; still keeping to wind'ard. And then at last they picked up the white bulk of the *Gil Eanes*, looming like an iceberg in the fog. Icebergs do not blow upon steam whistles: they knew that they were safe.

Their search for the ship presented a track roughly like this:

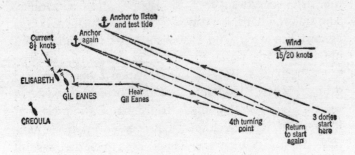

Some dorymen had been adrift for weeks and had survived. Two from the *Oliveirense*, a few voyages previously, had set off to sail for Nova Scotia when they found themselves hopelessly lost on the Grand Banks. They sailed and rowed for nearly ten days before a steamer picked them up and landed them at North Sydney. When I was in the *Gil Eanes* a doryman came aboard who had been adrift for five days, first in fog and then in hard wind. He had come back to his ship on the sixth morning prepared to get more bait and carry on with his work, but, to his surprise, he found that he could not walk, and was compelled to rest a few hours.

This man's name was Antonio Rodrigues Chalão. He was a slight, dark man with a shock of black hair and a clear, well-featured face, with grey-blue eyes and a straight, high brow. His home was at Vila Nova da Gaia, opposite Oporto, and he was a lifeboatman on the Douro Bar when he was not a doryman. He spoke

Antonio Rodrigues Chalão, Who Was Adrift Five Nights

The *Gil Eanes*

The Endless Job—Fixing the Long-Line

The *Elisabeth* Sailing from Greenland

The *Gazela* Got Under Way for Lisbon, Full—Going Aloft to Loose Sail

Their Ships All But Full, the Captains Enjoy a Cheerful "Gam"

Young Dorymen, Cheerful at the End of the Campaign

softly, with a quiet voice. His trousers were greatly patched and so were his brightly checked shirts. He wore an old jacket with the pockets torn, a rough cap, and high rubber sea boots issued by the company.

He had, he said, only a few cold beans in his pail when he left the schooner the morning he was lost—no bread. He did not usually take bread. He had a bottle of cold water. He went adrift in the fog because of an error in his compass. He searched and searched for his ship but could not find her. Then he anchored. He thought he was not far from the ship and, the next morning when the fog lifted, he saw her less than a mile away. But the wind and sea got up and the schooner was shut out in squalls of rain. He was driven off again and for the following four days could only battle to survive. He put his anchor down again, and rowed the dory to keep her head to the sea, bailing furiously every few moments to rid her of water. The weather was very bad, with a violent gale of wind, and he feared that his schooner would be driven from her anchorage and would not be in sight when the weather cleared. Day and night, without rest, he had to keep skilled control of his dory to see that she was not swamped. His experience in lifeboats was of value then, but he was ready for the end he thought inevitable. He thought of his wife and home and his seven children, and the fate in store for them if he did not return. And he redoubled his efforts to remain afloat, though the spume was flying and the wind shrieking round him, and the little red dory seemed hopelessly handicapped by the odds against her. He prayed, and the days passed, and the nights. He lost count of them. Surely some time the weather would let up. But still the wind screamed and the dory leapt and cavorted at the end of her grapnel line. This was nothing but a piece of thin rope but it held. He had already jettisoned much of his fish, keeping only a quintal or two as sailing ballast and as food in case he had to sail to Canada or Labrador. He pulled some folds of the sail over him to serve as shelter. In the nights he thought he would die of cold. On the third day he ate a piece of the neck of a cod: his fresh water was gone. But later the fog came down again and he could wring moisture from his woollen cap and drink that. It was very salty but it kept him going.

Then one morning—he did not know which morning; it seemed months since he had gone adrift—the fog all cleared and, when he was thrown on the crest of a sea, he saw the masts of a fishing schooner, afar off. He recovered his grapnel and rowed and sailed towards the schooner. When he came nearer, he saw that she was his own ship. Aboard they had given him up for dead.

Antonio Rodrigues Chalão came aboard the *Gil Eanes* to hear Mass in the little chapel down in the 'tweendecks hospital with the patients and the crew of the ship who were off duty. The patients who could get up sat in a 'thwartships row; the others—pale, bearded, gaunt—lay in their cots, following the service. Father António de Sá Rosa, on his first voyage to the Banks, had a large congregation, and many souls in his care.

Within a fortnight, the *Gil Eanes* had made the round of all the ships on Store Bank, finding something to do for all of them. Then she began to make the round again, more leisurely. On the first occasion, the sick and the mail were attended to. On the second, she delivered stores, which often included fresh water. I wondered how the ships had managed for fresh water in the days when there was no *Gil Eanes* to look after them, but it was explained that they used more water now for the cod-liver oil plants, and water was not quite so rigorously rationed when they knew they could get more.

The old ship wandered nearly to 70 degrees North, crossing and criss-crossing the Bank. When it blew hard, as it often did, she could not anchor, for she had already lost three anchors and dared not risk the loss of another lest she should not be able to anchor when the time came to touch at a port. She had plenty of the type of anchor the schooners needed, but no spares for herself. Anchors were important when it came to fetching-up a ship in small Greenland harbors like Sukkertoppen, where there was just room for the *Gil Eanes* and little else. When she had given away all the fresh water she could spare, she had to touch at a port to replenish. The *Santa Isabel,* the four-masted *Novos Mares,* the *Senhora da Saude,* and many more were in need of water, for some of them began to fear that they must go down to the Grand Banks again when the Greenland season was ended, and they did not wish to touch at St. John's if they could help it. Little Captain Alfredo Simões, of the *Santa Isabel*—the one master from the Algarve—had, however, no intention of going down to the Grand Banks. He was short of water because he had filled some of his tanks with bait.

"The *Gil Eanes* can give me water but she cannot give me bait," he declared. "And bait is vital."

The little captain—lithe, dynamic; his language explosive, unceasing, and to me quite unintelligible no matter whether he spoke Fuzeta Portuguese or his own version of Gloucester American—wore a sou'wester lined with a flannel cap on his grey head, and neither the sou'wester nor the flannel cap ever moved in any circumstances. His feet were sea-booted and his jacket was of stout pilot cloth, guaranteed to keep out the Arctic cold. He spoke at a furious rate of his early days fishing from the Algarve, of being a doryman out of Gloucester at the age of 16, and fishing the winters on Georges Bank, with his father and uncles, until the '14-'18 war called him back to Portugal. Alfredo Simões was the only ex-doryman from the Algarve then in command. He owed his advancement to his own energies and abilities, integrity, and intelligence; to the foresight and lack of prejudice of the Ramalheira, who recommended him; and the enterprise of the Bensaudes, who gave him his first command. A doryman from the Algarve—or anywhere else, except perhaps Ilhavo—had a hard road to the poop. Now his pretty schooner lay anchored a cable or so from the *Gil Eanes*, and she was within 2,000 quintals of full, as her captain confided in a conspiratorial whisper. He did not want much water. He was sorry about the bait, but what could he do? Fish he *must* have: give him three weeks of even average weather and he'd be homeward-bound.

Homeward-bound. It was the first time those words had been uttered since we had left Portugal. All the other masters were still full of bewailing, except those of the seven schooners on the Grand Banks. Their news was passed to the *Gil Eanes* by telegraphy through trawlers on the Banks when the radio telephone could not bridge the distance. The seven schooners there were reported to be doing very well, for by mid-July the cod were in abundance round the Virgin Rocks, and the little *Lousado* of 220 tons, the baby of them all, with only twenty-eight dorymen, was reported to be nearly full already. Even the unhandsome *Maria Frederico* had taken 2,000 quintals in a week and looked like going home soon, a full ship, though her master had said nothing of his hopes of being homeward-bound.

"When the *Maria Frederico* fills, there are plenty of fish," Captain Vitorino said. Many masters on the Greenland grounds began to declare that they would hurry to the Virgin Rocks: a dozen more were clamouring for water.

"We will go into Holsteinsborg and get them some," said Commander Tavares de Almeida, "and the sooner the better. It is a lot of work for the *Gil Eanes*, for there are no tanks. But go we must."

All the parcel mail and stores had been delivered then and the odd jobs done. The sick and ailing had been rounded up, transferred when necessary, and treated in their own ships when that could be done. For once the weather was good and the visibility excellent. The old steamer stood in towards the wild mountains of Greenland and in a few hours the hamlet of Holsteinsborg showed in a gap between the rocky islands. The approach was dangerous but well marked, with leading marks to guide vessels in from the south. Once or twice the *Gil Eanes* was very close to horrible outcrops of rock, but she came in safely to anchor in a pleasant cove surrounded by rocky islets and mountains, bold and bare. The steep-roofed houses of the little port were perched on rocks round a tiny inner cove, where a horde of hungry husky dogs were tearing at what remained of the carcass of a whale. At the stone jetty some Eskimo and part-Eskimo children were fishing nonchalantly, and the little harbour was so full of boats that it was almost impossible to approach the landing-steps. The Danish flag flew above the Resident's house on the hill near the school, and clouds of hungry mosquitoes threatened to make life miserable.

CHAPTER FIFTEEN

THE GRAND BANKS AGAIN?

> Black was the Sea, and at long distance bray'd
> As if it roar'd through Rocks, down Rocks did fall.

IT TOOK the *Gil Eanes* four days to embark a hundred tons of fresh water at Holsteinsborg, for it all had to be taken from a mountain stream and ferried to the ship in barrels. The watering place was on the northern side of the outer harbour, and the ship's lifeboats and wine barrels were used for the job, which was made easier by hiring a small motor-boat to do the towing. The mosquitoes at the watering-place were insufferable. It was easy to see why the Eskimos pulled big hoods over their necks and ears. The mosquitoes flew into the sailors' eyes, ears, nostrils, and mouths, and it was of little use to brush them away, for always more came.

The weather on that side of the harbour was quiet and sunny, but the mosquitoes ruined what might otherwise have been pleasant days. There was no ice in the harbour, and very little snow on the hills near-by. In the little port, most of the local inhabitants, called Greenlanders there because of the considerable mixture with the original Eskimo blood which several centuries of whaling and trading had brought, went about in clothing little different from that usually worn in milder countries, except that they all wore ankle moccasins, made of sealskin. Some of the older women still wore sealskin trousers and decorated knee boots, but many of the younger women might have passed for high school girls in Honolulu, except for their feet. Most of the houses were carefully sited on outcrops of rock where the winter's snow could neither dislodge nor bury them, and it was astonishing that the children and dogs playing about did not tumble down the precipitous cliffs, a step or two beyond their back—and front—doors.

There was not an Eskimo kayak or even the remains of one to be seen at Holsteinsborg, but several interesting little vessels were in port. These included several small fishing cutters with the typical double-ended North Sea fishing hull. Some of these were less than forty feet long but they had all come under their own sail from Denmark, across the North Atlantic and up Davis Straits. Such passages were commonplace, and no one paid any attention to them. The prettiest of these vessels was a lovely little blue yawl which was used for surveying and for fisheries research. The success of the cod fisheries in coastal waters had become of vital importance to the Greenlanders, and the Danish government maintained a biologist who, working from Copenhagen, spent the summers in Greenland waters marking and examining cod.

This biologist, at the time of our visit to Holsteinsborg, was Dr. Paul Hansen, who had been one of the

Danish delegates to the Cod Conservation Conference at Washington, D.C., when Commander Tavares de Almeida was one of the Portuguese delegates. Dr. Hansen, unfortunately, was not at Holsteinsborg while we were there: he was away marking fish somewhere in the neighbourhood of one of the more important spawning fjords. He spoke to the ship daily on the radio telephone. The *Gil Eanes* had collected from the fishing schooners a number of biologists' tags and other marks from fish—Norwegian, Danish, and one from Newfoundland—which were of interest to Dr. Hansen. Though the Greenland Administration had carried on continuous investigations for many years, there was still a great deal to be learned about cod in those waters. Were they being over-fished? Would the too thorough modern methods bring about such a decline in stocks that it might become unprofitable to fish for cod any longer? Could good and bad seasons be predicted? They ought to be. After all, the conditions in which cod thrived were fairly well understood. Apparently, whether there were cod in abundance during any season or not depended to a great extent on the numbers and size of the year-class which should come to maturity in that year. So long as the temperature and salinity of the water remained right, and the continuance of the warm cycle encouraged the growth of food for the cod, the cod should thrive. But some fry got off to a better start than others. When there was a season of good conditions for the fry there were certain to be cod in plenty on the Banks some years later, when those fry had had the chance to reach maturity—always provided the warm cycle continued.

Dr. Hansen discussed this and other uncertainties. It

was his opinion that the present warm cycle, which had been continuous since the 1920's, showed no sign of passing, and that it had brought many changes to Greenland waters besides the abundance of good cod. Nobody knew where all the Davis Straits cod spawned: some certainly went to Iceland, for cod marked by biologists off Greenland had been caught in Icelandic waters. One had been taken as far away as the Barents Sea. Investigations were being carried on from Greenland, Iceland, Norway, the Faroes, Newfoundland, Nova Scotia, and the U.S.A.

If the cod did move away from Davis Straits, no one knew what the Greenlanders would eat, for they were now accustomed to taking thousands of tons of cod each year—in 1942, 10,000 tons; in 1946, 13,000 tons. There were over a hundred curing houses established along the west coast, and practically every coastal Eskimo took some part in the fishery. When the old Greenland brig *Tjalfe*—which for so many years later was a prominent feature of the lovely Copenhagen waterfront—was investigating the fisheries of Davis Straits in the first decade of the century, cod were rare. They might very easily become rare again but the good doctor promised an excellent year-class in 1955.

This information, interesting though it was, did not help the dorymen and the captains on the banks in 1950, and the *Gil Eanes*' radio continued to broadcast news of bad weather and poor fishing. It was blowing fresh outside during all the four days we remained at Holsteinsborg. The *Senhora da Saude* announced that she had had enough of it and was shifting down to Fyllas Bank to see if the weather was better there. There were thirty-eight Portuguese line-fishermen still on

THE GRAND BANKS AGAIN?

Store Bank, and the captains of all of them were restless. They were more restless when they heard of the success of the schooners which had remained on the Grand Banks, for by early August one at least of these announced that she was ready to leave. News of this came through the *Gil Eanes*. Everyone was pleased, of course, that the smaller schooners were for once doing very well, but the pleasure was tempered by their own misfortunes.

The news from the Grand Banks was not all good, for the *Lousado* lost a man, a day or two before she was due to sail, and the little three-masted schooner *Ana Primeiro*, of Figueira da Foz, just as she had filled with fish, burned out and sank. She was one of the older-type schooners with the galley in the rancho; and the combination of open-fire stove and open barrels of olive oil had more than once led to the loss by fire of a tinder-dry wooden schooner. Apparently it was not the galley which had caused this loss. The fire was reported to have been caused by an electrical short circuit. It had a good hold before it was discovered: the *Ana Primeiro* was doomed. She was a typical Baltic schooner, built as the *Erika* at Ornavik in Sweden in 1918. She had a Scandinavian bow, a square stern, and a sea-kindly hull. She was of 275 tons. It was most unfortunate that she should have burned out just as she had filled with fish for she had had poor luck the previous year and needed a successful cruise. Her people, taking to their dories, were all saved by other Portuguese schooners, and there was no loss of life. She had been considerably battered by the hurricanes of 1949, and her re-fit for the 1950 season had been thorough and expensive.

Her loss was accepted philosophically: the great

thing was that her people were saved. The Bankers were used to the losses of old schooners—perhaps one, perhaps two—during most seasons. Fire took some; others opened up in the giant North Atlantic seas and foundered. Very few were lost navigationally. The *Ana Primeiro,* however, had narrowly escaped being driven ashore on the Newfoundland coast during the previous season, when a re-curving hurricane caught the seven Grand Banks schooners as they were trying to make for St. John's. With sails blown away, decks gutted, some with the helmsman lashed to the wheel to prevent his being washed overboard, the seven small schooners struggled towards St. John's through the murk and noise and violent seas of the hurricane. That was a test of ships and men. The *Maria Frederico* lost her rudder; the *Navegante II* had opened up and foundered at the hurricane's first passing across the Banks, and her people were in the *Maria Frederico.*

That night the radio telephone of the *Gil Eanes* had spoken quietly of the approach of tragedy, for the seven schooners got off course, driven by wind and sea, and the radar showed them to be driving steadily towards the rocks of the harbour entrance. The little *Ana Primeiro* had extricated herself, by her own supreme seaworthiness and a feat of brilliant seamanship on the part of her young master. Her compass gone, her charts reduced to pulp by a sea which broke through the saloon skylight, with only the close-reefed peak of the fores'l and a cloth or two of the fore stays'l set, she drove towards St. John's. Off the harbour mouth, the *Gil Eanes* picked her up through the port-radar, and guided her in by the radio telephone, as if she were an aircraft coming in to land blind with only the radio telephone

THE GRAND BANKS AGAIN?

to guide her. That was a night. Then the hurricane had curved again, this time away from the land and, in the nick of time, the other schooners cleared the lee shore. If they had not, if the wind had not changed just then, it is most unlikely that any schooner or any fisherman would have survived.

There had been many tragedies of that sort in the past. In the past ten years more than a dozen schooners had gone, not counting the *Maria da Gloria* and the *De Laes*, lost in the war. The *Silvina*, *Ernani*, and *Julia IV* had all succumbed to fires originating in the galley. The *Gaspar*, *Santa Quiteria*, the barquentine *Normandie*, the three-masted *Maria Carlota* and the *Navegante II*, deeply laden in the autumn months, had all foundered. They were small ships, and it was a hard trade. Salt was a heavy dead cargo, and so was salt cod. Spending months at anchor on the turbulent Banks was a strain on any timbers and, almost from 1918 onwards, it had been difficult to find good wood to build into ships. No wonder, then, that a schooner occasionally opened up, and the inrush of the wild North Atlantic was more than her pumps could cope with. There was, for example, the *Maria Carlota* which, homeward-bound deeply laden with cod one October, was swept by storm after storm until she became water-logged, in peril of sinking. A liner, New York-bound, took off her people and from New York they were flown back to Portugal, overnight, in a chartered aircraft. It was an odd experience for the dorymen.

Schooners could be lost on the very doorstep of home. Many of the fishing ports were dangerous to approach in heavy weather. At Oporto, Figueira da Foz, and Aveiro, there were dangerous bars. Both the *Neptuno* and the *Primeiro Navigante* coming home with full cargoes of fish had run upon Aveiro bar. The *Neptuno* broke her back and was got off again, to sail no more; the *Primeiro Navigante* stayed on the bar.

As soon as her watering was finished, the *Gil Eanes* hurried out from Holsteinsborg. One of her last visitors was the surgeon from the Danish hospital ashore, to talk about the recovery of a Portuguese fisherman whose leg he had had to amputate the previous season in an emergency, and to look at the serious cases in the *Gil Eanes'* hospital. He was a quiet, pleasant fellow, about to go down to Copenhagen for a short leave after several years in Greenland. He made his rounds by dog-sled most of the year, he said, and he liked the Holsteinsborg district. The mosquitoes were bad for two months only of the year. Two months! Usually there was a fine spring, bright and sunny: the Arctic night lasted little more than a month though they were beyond the Circle. On the whole, he declared, the climate was not too bad.

Big dog-sleds leaned against most of the houses, and huskies were asleep in the shade, dirty, bad-tempered and bored. The huskies seemed either to be asleep, trying to catch fish in the harbour, or fighting. The local Greenlanders did not believe in feeding them during the summer months, when they did no work, though there were pieces of whale-meat hanging on nails outside most of the houses, to provide them with food. The huskies were good fishers, prowling in the shallow water at the head of the cove on the look-out for stray cod. They had torn the carcass of the stranded whale to pieces and there was no food left on it.

As the *Gil Eanes* steamed to sea and the brightly

THE GRAND BANKS AGAIN?

painted little houses grew smaller in the distance, I wondered again why the children and the dogs did not come plunging down the cliffs on which they were at play. Perhaps some did. There has been a considerable increase in the number of houses at Holsteinsborg during recent years and there was much building activity while we were there, on workshops, waterworks, and new homes. There was a shrimp cannery, but the shrimps had chosen to desert the area and now had to be brought from miles away. This was another problem for Dr. Hansen, the biologist. Why did the shrimps go? Some of the shrimps had certainly gone into cods' stomachs for, aboard the *Argus* and the other ships, many cod were stuffed with small shrimps or prawns.

Early in August I was aboard the *Argus* again. By that time several trawlers had arrived on Store Bank, including a large Italian, several Spaniards, and a Scot named the *Loch Inver*. There were claims against the Italian for the destruction of dorymen's lines, but her master asked plaintively how he was to find room to trawl when there were 2,000 dorymen on the bank. The French were able to do it; they rarely impeded the dorymen. There was room for all. The schooner captains were not much worried about the trawlers, which for the most part kept to the outer area of the bank. The hand-liners fished the middle and towards the Greenland coast outside territorial limits. What they needed was tolerably good weather.

The cod were still lance-crazy and most of the captains were still pessimistic, though in unguarded moments a few had been heard to say that they did not intend to go again to the Grand Bank. This was taken to indicate that their fishing was not so bad as they

THE QUEST OF THE SCHOONER ARGUS

made it out to be. Among these captains was Captain Adolfo, though he still declared that his fishing was very poor. He would, he said, fish in Greenland waters until mid-September at latest; then, full or not, he would sail directly back to Portugal. A day or two later, Captain Silvio was actually heard to offer bait—to give mackerel which he did not require—to other ships, and that day his radio blared marches, tangos, and fados, for hour after hour. There was at least one happy captain on Store Bank, then. But Silvio still declared that the *Elisabeth* was not full, far from it.

"A fishing-ship," Captain Adolfo said, "is never full. You can always find room for just one more cod. He'll sail when he has used up his salt and has 12,000 quintals stowed below."

The *Argus* herself was deep in the water, much weather-stained and marked by the dories where they had rubbed alongside. Her sleek sides were not so white as they used to be, and the grime from the oil plant disfigured some of the masts and rigging. The cold water washed across her main-deck though she was anchored, for there was a big sea running. I rejoined her by lifeboat, and the 26-foot boat seemed large and supremely safe after the dories. I had to jump for it at the rail. Captain Adolfo, César Mauricio, the mate João Matias, the second mate José Nunes de Oliveira, and the assistant engineer Manuel da Maia Rocha all had luxuriant black beards. They were tired but they looked well. Only Captain Adolfo seemed pessimistic, but I had begun to think that, in his case, pessimism had been a pose for so long and was so much a part of the deep-rooted tradition of his calling, that he could not be optimistic if he tried. Each member of the afterguard made

THE GRAND BANKS AGAIN?

his personal estimate of the ship's fish. César said his total was nine and a half thousand quintals. The mate said just under nine thousand, and the second, just over. Captain Adolfo said nothing, and I looked in the hold. The fish were well stowed in all but two of the twenty-seven compartments: these two were still full of salt, and very small. The dorymen said another three thousand would fill her, if only the weather would give them a chance. But it was still blowing fresh and the sea was high. The dorymen were busy stowing their long-lines away again, for Captain Adolfo said he would waste no more bait on fish which refused to go to the bottom to look for it. On the morrow they would jig. They had tried the long-line only for a few days but it was hopeless.

It was evening, and the golden light from the setting sun made the coast of Greenland beautiful. About Isortok the higher hills were snow-covered again, after having been almost bare for several weeks in midsummer. The entrance to the Nordre Stromfjord (locally called Nagsugtok) showed clearly from the sea though it was a good twelve miles away; but the entrance to the nearer Isortok remained obscured. The sea was going down quickly: the weather might be good for fishing on the morrow though there were heavy banks of cloud in the south. About a dozen Portuguese hand-liners were in sight from the poop of the *Argus;* and a couple of big Frenchmen, so miraged that they looked like *Queen Marys,* were steadily trawling away, out towards the seaward edge of the Bank. Through the loudspeaker on deck came the voice of little Captain Alfredo Simões of the *Santa Isabel*, expressing his fear that the weather would not allow of good fishing. Moreover he was short-handed. He had three dorymen laid up aboard, he said, and another two in the *Gil Eanes*.

"Yes, and he's putting fish in the freshwater tanks now, I wager," said the mate. Perhaps he was right, for the *Santa Isabel* had looked well-laden when last she was near the *Gil Eanes*.

The next day the *Argus's* dorymen jigged more than 150 quintals of fish. Some of the men were fishing with lance bait which they took fresh from the stomachs of jigged cod. They used one jigger and one hand-line with bait but, with or without bait, it was very hard to fill a dory. The older men came back tired, their faces drawn and grey, though their eyes were still bright and alert. Hauling a jig-line and a hand-line all day, going back and forth over the same stretch of turbulent water, was gruelling work. Abilio, the Boss of Salt, who had been laid up six weeks in the *Gil Eanes*, was back again and out in his dory. Abilio was fifty years old and had suffered badly from bronchitis. Captain Adolfo had suggested that he should remain in the Algarve that season for the good of his health, but he preferred to make the voyage. So there he was, up from a sick-bed and away for a fourteen-hour day of jig-fishing, the water near thirty fathoms deep and the fish not plentiful. Nevertheless he filled his dory and looked pleased as he came back and gaffed up the heavy fish. Then he superintended the salting, which Captain Adolfo had been doing in his absence.

The following day the dorymen jigged 140 quintals. The *Argus* was drifting with them; and near-by, the *Gazela*, the *Coimbra*, and the *Condestavel* were drifting too, with crowds of dories all round them. That night a fog came up suddenly with a breeze of wind,

THE GRAND BANKS AGAIN?

and again dorymen were missing—thirty or forty of them from schooners elsewhere on the bank, which had been long-lining. Fifteen were adrift from the *Maria das Flores*. By midnight none had been found. It was pitch-black and cold and miserable. Then the fog cleared, and one of the *Maria das Flores'* dorymen returned to the ship, to report that the others had run in ashore and landed in a fjord. It was the only safe thing they could do, for the tide was very strong and setting them away from the ship. The sea was getting up dangerously, and they all had good fish. Over the coast the visibility was good, and they had soon found a lee. They landed through kelp, made a fire, cooked a cod in the ashes, and settled themselves for the night. When the tide turned they sent back the most lightly-laden dory to find the ship and report where they were, but it was past midnight before this dory found the schooner. In the morning the local Eskimos found them on the beach and made them welcome. While they waited for the weather to moderate—it had worsened again—the Eskimos trimmed their beards and hair. So the dorymen came back to the *Maria das Flores* in better shape than they had left, and they knew that they were fortunate.

On the night of the 20th of August, the little schooner *Rio Lima*—a three-master from Viana do Castelo—suddenly announced that she was full and would be sailing for Portugal on the morrow. This was a fortnight after the first schooner had sailed full from the Grand Banks, but it was very good. The *Rio Lima* had thirty-eight dorymen and a capacity of approximately 7,000 quintals of well-stowed fish. It was good to know that somebody was full on the Greenland grounds. We had not heard any rumour that the *Rio Lima* was filling or doing particularly well. Indeed, rumour—and the dorymen's news exchange—had mentioned several other ships as being likely to sail first from Davis Straits: the *Condestavel,* for instance, and the little motor-ship *Terra Nova,* or Captain Silvio's *Elisabeth.* The Viana ships kept to themselves, though their captains were from Ilhavo. We did not get much news from them, and we rarely found ourselves fishing anywhere near them. They had been the first to sail from Portugal, immediately after the Blessing. It was their immoderate haste to be gone which had sent the whole fleet to sea much earlier than usual. Mid-April would have been time enough to sail, as the event proved.

The next day, fish were put in the last compartments aboard the *Argus.* For the sake of speed, fish were stowed in these compartments, which were clear and easy to work in, as they came down the chutes cleaned from the vats. On the following day, if the weather was bad and the dories could not be launched, the salters went below again and restowed much of the fish in other compartments, close below the deckhead, where it was extremely awkward and trying for them to work. The salters all wore rubber gloves because their hands were split and torn by the sharp lance which they used for bait. They lashed rubber pads to their knees and put fresh lashings on their oilskins, but the salt and the cod penetrated everywhere. They had to work on their backs, taking care with each individual fish. By night, as the fish mounted in the last compartments, the salters' work became very difficult. Salt had to be passed down to them from the boxes on deck, and this kept a gang of men busy. They had to work in a space which be-

THE GRAND BANKS AGAIN?

came more and more constricted, but every day the cod continued to shrink and the milky pickle was pumped away twice to make room for more and yet more fish. Time and time again they filled compartments to the very beams, only to find room for more fish there a few days later. It seemed that we should never fill. There was still plenty of salt and there was still enough bait, because of the saving caused by the use of the jiggers. But the dorymen became more cheerful. They, like Captain Adolfo, had decided that this year there was to be no second visit to the Grand Banks.

The sailing of the *Rio Lima* caused some suppressed excitement but, for two days afterwards, bad weather kept our dories nested. Ships to the south gave warning of the approach of this bad weather and no dories were lost. One advantage of the use of jiggers and handlines was that such fishing could be stopped instantly and, if bad weather came, the dories were at once able to return to their ships. Towards the end of August the bad weather became much worse and on many days it was not safe to launch the dories, even for a few hours of jigging.

Meanwhile the *Lousado* had arrived at Lisbon. And after the *Rio Lima*, the *Condestavel* announced that she was full and about to sail, and then the *Elisabeth*. Captain Silvio had taken a big stock of bait at North Sydney, and had fished very successfully with the jiggers at lance-time. He had given us the bait which the *Elisabeth* had not needed. The *Argus* was using the long-lines again—one day she took nearly 300 quintals—and could do with it. Other ships were still jigging when Captain Adolfo decided to go to the middle of the bank and try the bottom again. It was a good move.

THE QUEST OF THE SCHOONER ARGUS

That day the *Elisabeth* came round our counter to wave farewells. It was evening when she came, and her fishermen were cleaning fish silently by the pounds. Captain Silvio was on his bridge, with a fur trapper's cap on his head and a leather coat draped about his shoulders. A few fishermen looked up and waved; the Captain and the mates waved too. The little *Elisabeth*, very deeply-laden, dipped and rolled and wallowed off towards the south. She was to embark a surgeon and some seriously ill men from the *Gil Eanes*, to get them quickly back to Portugal. The *Condestavel* had taken the walking cases with a nurse. The sick were always sent back as quickly as possible. The *Gil Eanes* was looking after the schooners which had moved down to Fyllas Bank and it would be a long time before she was back in Portugal.

The *Elisabeth* carried fifty-six dorymen and had a capacity of not more than 12,000 quintals, so that her dorymen were not required to make the high average catch demanded of the *Argus's* men. The *Argus* had only fifty-three dorymen and could hold more than 13,000 quintals. The *Elisabeth* also profited by the astonishing luck of Captain Silvio. She had, they said, never been fished full by any other captain. The *Condestavel*, too, carried fifty-six dorymen for a capacity of 12,000 quintals. It was my guess that the *Argus* had little short of 12,000 quintals aboard at that time, though no one would admit it, and the official total given out by Captain Adolfo was not much more than 10,000. Captain Adolfo was obviously doing his best to stow the ship as full as possible. Nobody seemed to know where the Plimsoll line was: nobody knew and nobody looked to see. There was no use looking. The

THE GRAND BANKS AGAIN?

ship's hull midships where the load-line was cut was covered with grass and it was impossible to see how deeply she was loaded. Not that anyone worried about that. If a ship did not leave the grounds at least a little over-laden she would never arrive in the Tagus with a full cargo. It was sure to be shaken down unmercifully on the way across, and the pumped out pickle relieved the ship of a good many tons.

"She has sailed home with 14,000 quintals," Captain Adolfo said. He did not disguise the fact that he would like nothing better than to do so again.

"And no dories," said the mate, for they had all been washed away.

Captain Adolfo grinned. Dories could easily be replaced, but cod were gold. The *Argus* was a very strong ship. It was obvious that she needed to be.

By Monday, the 28th of August, the cod-liver oil production was so good that two dozen large oil-drums of the stuff were stowed on deck, and the engineers began to run the oil into the big forepeak tank. The cod-liver oil tank itself was already full, and every spare drum and barrel in the ship was filled or being filled with something—cods' cheeks, tongues, backbone membranes. The deckboys began to clean out the spare dories which had held salt, and to stow salted cods' cheeks in them.

But that day the long-lines yielded a miserable hundred quintals. On the next day there was no fishing, though the Italian trawler passed close by, dragging his great trawl. Our crowd of black-marked seagulls, raucous and hungry, stared at the ship in astonishment, wondering why she was throwing them no food. When fish-cleaning was in progress they feasted magnificently;

and they loved to gather round the schooners with oil plants because the bits of liver in the boiler debris were great delicacies. The older schooners had no such plants, and consequently few gulls.

When the *Argus* had no fish, the gulls, after waiting about hopefully for some time, would fly disconsolately away in search of a trawler. The gulls we saw were mostly small and white, with black face markings like Shropshire sheep.

The second mate kept an injured one in his cabin as a pet. When it recovered, it flew away. There was no attempt to catch and use them, or any other birds, as bait.

The last three days of August gave us less than 150 quintals, and once again there was a fear, fore and aft, that we might be compelled to return to the Newfoundland Banks. The schooner *Antonio Ribau*, the little motor-ship *Terra Nova*, Captain Alfredo Simões' *Santa Isabel* and the Viana four-master *Santa Maria Manuela*, had all sailed for Portugal, and the *Coimbra*, *Dom Deniz*, and *Senhora da Saude* were reported to be ready. On the 30th, the *Gil Eanes* left the Greenland grounds to return to the Grand Banks and she came across to the *Argus* on her way. Commander Tavares de Almeida, a grey-haired stalwart figure, waved from the bridge, happy that after all the alarms, not a doryman had been drowned in Davis Straits this year and, as yet, not one had died in the hospital ship. The Commander was a man who felt very deeply for the dorymen and for all who were concerned in the Arctic fishery. So long as a doryman was fishing in Davis Straits, the Commander knew no peace. When, as so often happened, men were adrift by the dozen, his anguish was

THE GRAND BANKS AGAIN?

as deep as that of any master. In a Davis Straits season half-a-dozen dorymen usually failed to return. That was the average. There were sometimes more.

Belching smoke, the *Gil Eanes* departed. We were sorry to see her go. For the *Argus*, August ended as it had begun, with poor weather and miserable fishing. The dorymen could not average a quintal apiece on the last day of the month. The ship had not taken 3,000 quintals during the whole month. It was the worst August she had known in twelve Greenland voyages. Should we, after all, have to return to the Grand Banks? There would not be many more days on which the dories could fish off Greenland.

CHAPTER SIXTEEN

END OF THE CAMPAIGN

> The men are harast, and with Travaile broke,
> 'Tis now high time (as it appears to me)
> To shew them that Land where they would be.

"A DAY of excellent weather—for once—but yet again of poor fishing. Where *are* the blasted codfish? We get perhaps a hundred quintals, which, for a good day in September, will never do. There won't be many good days in September. The fifty-three dorymen fished a fourteen-hour day, with jigs and long-lines, but even the redoubtable First Fisher could not fill his dory, though he tried every way he knows." So begins the entry in my diary for September 1, 1950. We were well off the coast, not far from the island of Umanak, towards the north of Store Bank, and there were many other Portuguese ships still in company. For several days the *Argus* had been trying her luck off Isortok but did so poorly that she had moved to the Umanak area. That very day the *Lutador*, which had just shifted there, fished very well off Isortok. A chancy business! Our

END OF THE CAMPAIGN

night watch had jigged up a quintal or so of very small cod, not much more than a foot long. When he heard this, Captain Adolfo weighed and moved off, for it was certain that where the small cod were feeding the big cod would not be found. The big cod are cannibals, and the smaller fish keep away from them. Small cod salt down to skin and bones and are useless as preserved food. They were of no use to the *Argus*.

It was all most disappointing. One week's good fishing would give us a good cargo but the schooners could remain on the Greenland grounds only another fortnight at the most. Much anxious scanning of the *Argus's* previous logs showed that whenever she had been compelled to remain there into September, she had enjoyed few fishing days. Bad weather, bad weather, bad weather. And hopeless to launch the dories. Out of ten days in September 1949 she had been able to fish on only two: the best September she had known there had allowed the dories to be launched no oftener than one day in three, and then not always for the whole day.

The fishermen hated the idea of moving down to the Grand Banks again for it was sure to add another month to the voyage. Already it was late to be fishing anywhere, either on the Grand Banks or off Greenland. Off Newfoundland, September was a hurricane month. It would take at least ten days to sail down to the Atlantic grounds, and then what? In thirty days of a second visit to the Grand Banks the year before, the *Argus* had been able to fish on six, and had taken less than a thousand quintals. Hanging on at anchorage day after day in savage weather or, worse still, pounding about under short canvas unable to anchor at all; crossing the liner routes in sou'west murk and blinding rain with a gale of wind blowing which never seemed to end; watching the dories washed overboard and the fish-trolleys smashed by the breaking seas—this sort of thing was trying at the end of a long season. The dories had to be nested on the exposed main deck, for there was nowhere else for them. If more than a certain number of them were washed away or smashed the fishing must be abandoned.

Yet that day, three of the hand-liners, despairing of ever fishing full in Davis Straits, had lashed their dories and left to sail down to Newfoundland. The *Lousado* and all the other smaller schooners had arrived at Lisbon long ago. The captains of the hand-liners hoped to find the fish which had filled the small schooners so handsomely, and off they went—the motor-ship *João Costa* (which had been trying the banks off Disko), and the three-masters *Infante de Sagres* and *Labrador*. The *Labrador* had been fishing very poorly, partly because her dorymen were unaccustomed to the long-line which they were using that season for the first time. All dorymen hated innovations. The *Labrador* was a schooner of just over 300 tons and she had always been filled before by jigging and single-hook hand-lining, but that season she was fitted with a refrigerated chamber for bait and could use long-lines.

As soon as our dories were all back and nested, we weighed, and off we went again on the eternal quest for cod. But where should we go? Why should any spot on the banks be better than another? Captain Adolfo was a knowledgeable man about the Store Banks fishing, and had spent as much time there as any schoonerman: he had, perhaps, a trick or two up his sleeve yet. We would shift anchorage, then jig from the poop for

END OF THE CAMPAIGN

an hour or so to test the ground and, if the cod brought up were too small or none were caught at all—as often happened—we would shift again. There were some places where the bottom was comparatively richer in food for cod, places where a confluence of the tides or a meeting of waters coming out from the deep fjords created conditions unusually favourable to crustaceans, or to shrimps, or caplin or anything else which the cod could eat, including the young of its own kind. Captain Adolfo knew these places and we hunted on them.

But the trawlers were beginning to assemble in large numbers on the Davis Straits grounds. There were already thirty or more about, all enormous, and all trawling a twenty-hour day. We could hear them on the radio, discussing their catches. Sometimes they did well and sometimes they did not. They could trawl on days when we could not launch dories. Sometimes they came close by, gathering up the fish and sweeping all the codfood from the bottom. The Italian *Genepesca I* and a couple of huge Frenchmen, looking to us like 10,000-ton tankers, were close enough for us to read their names, and the *Genepesca* took away half the lines of the Little King. We had not a doryman who could better spare them, but his laments were loud and bitter. The Little King was trying hard, within his limitations, for he wished to visit the Grand Banks again no more than any other man.

Whenever it was safe at all, Captain Adolfo launched the dories, and the dorymen were always ready to go. On the second of September we took 150 quintals, and it would have been two hundred if the sea had been reasonable. Laurencinha had to dump fish for his life to keep his dory afloat, and came nearer to being drowned that day than he had been for some time. Other dorymen had to give fish away to dories from the *Antonio Coutinho*. The fishing was good but the dorymen dared not fill their dories, lest they should swamp and founder. By noon we had had six fair loads back; by one o'clock in the afternoon, a dozen. But a little after two, the recall had to be hoisted. It was too bad.

On days like this, when we managed to fish 150 quintals or more, the fishermen were all smiles; but the next day the dories had to stay in their nests and all was gloom again. The departures to the Grand Banks continued: only ships very near full were staying in Greenland. Old Marques, with the *Capitão Ferreira*, who had been trying his luck on Fyllas, gave up and sailed for St. John's to get more bait before trying the Grand Banks again. The motor-ships *Cova da Iria* and *Inacio Cunha* sailed for Portugal with good cargoes. The big *Milena*, the *José Alberto*, the schooners *Aviz*, *Brites*, *Viriato*, *Rio Caima*, and all the big 70-dory motor-ships —which were hard to fill—sailed for Newfoundland. Soon only the fleet of the Bensaude Line, the pretty four-masters *Novos Mares* and *Groenlandia*, the *Oliveirense*, and the motor-ship *Antonio Coutinho*, remained on Store Bank. In all, sixteen ships had left for the grounds off Newfoundland, whence the annual onslaught of Autumn hurricanes was already being reported.

Captain Adolfo stayed on. So did his brother Almeida, and young Leite in the old *Gazela*, and O Anjo of *Hortense*. Various preparations were being made in a quiet way for the homeward voyage—preserving cods' cheeks wherever a place could be found to stow them, putting down about a hundred fine cod in the port refrigerated room, and salting the flesh of a few repulsive catfish. The catfish were apparently regarded as a delicacy in Ilhavo. The fresh cod in the freezing chamber were intended as gifts to friends of the ship at Ponta Delgada, and as fresh food for the passage. Some halibut and flounder were also put down. We were still eating cod in some form or other twice a day; I was pleased—and not a little surprised—to see what excellent friends the afterguard remained after five and a half months of sharing two meals of cod every day. Below, the good captain, who was a very gentle fellow at heart, abandoned his pose of toughness. He ate little and rarely did more than taste the cod which, after 31 years, he abominated. The younger men still ate prodigiously but the older did not. Their conversation was always animated and friendly but the radio rarely allowed it to go on for very long without interruption. The radio was sacrosanct. It was permitted to interrupt everything, and everybody listened whenever a voice issued from it, which was practically all the time.

It seemed to be not so much what was said as what was not said, that was regarded as important; but on rare occasions a captain about to leave the area would give information that was really valuable. There were one or two captains who spoke little but always to the point, the pleasant Manoel da Silva in the *Lutador,* for instance. Then telegrams to owners and the Gremio had to be dictated by radio telephone to the *Gil Eanes,* or to other ships with telegraphy, and we could often gain considerable information from these. But some captains went to infinite trouble to disguise the state of their fishing, even when dictating a telegram. "Ask

Commander Tavares de Almeida to add 3,000 to the figure I gave him," they would say, or something like that, instead of stating their catch. The other captains could only guess. Their guesses, on the whole, were remarkably good.

Soon the four ships of the Bensaude fleet were together in company—*Argus* and *Creoula, Gazela, Hortense*. The *Creoula's* male nurse looked after the health of all the men in the ships still on Store Bank, and Captain Adolfo distributed salt and engine spares and whatever was necessary to keep the fleet going.

"Whenever you see the *Argus*, you all want something," he said to young Leite when the *Gazela* came over for a gam,* and to borrow salt and one or two things. The *Gazela* was jigging again, being out of bait, but she was so near to full that her dorymen were jigging through the long days with even fiercer energy and determination than usual. The cod were there again and they were catching them, except on patches where the trawlers had passed. The talk between the two captains was exclusively of cod, with much emphatic gesticulation and some shouting. I asked young Leite if he dreamed about cod, but he said not. He didn't dream at all. Nor did he re-live his campaigns in the cafes at Ilhavo. But on the grounds there was nothing but cod to talk about. Captain Adolfo's estimate of his take of fish was so obviously unrealistic—in a vessel clearly overladen, though she still had a good deal of salt—that Captain Leite jumped into the fish-hold to see for himself.

"Why!" he said when he had taken a hasty look down there, "she couldn't take another 700."

---

* An old sea word for a meeting of ships, with visits exchanged by boat.

Down Davis Straits

The End of the Voyage, Awaiting the Next

The Dorymen Went Back to Coastal Fishing

END OF THE CAMPAIGN

"Another thousand might give us a tolerable cargo," said the unrepentant Adolfo, with the water lapping at his cabin ports. "She's full of salt."

The mate confided afterwards that if we took 500 quintals in two days it would be "too fast," for the ship would be full before the last of the cargo had settled. It did not take long for the cod to thin down when the pickle was pumped off twice daily, though the fish were not salted too heavily. The dorymen hoped she would take 500 quintals on the morrow, "too fast" or not. So late in the season it couldn't be too fast. Most of the fishermen were showing strain. Just when the weather was really worsening they were called upon for extreme endurance and effort.

But the next day we took no fish, and the gale howled. The *Argus* was shaking her fish down with uncalled-for thoroughness: several compartments which had already had their "final" stowages twice now gaped with room for more fish. More fish!

"Schooners grow in September," Captain Adolfo said. "There are always holes for more fish. So I am always miserable." But he grinned.

Sometimes the afterguard's pose of pessimism was almost humourous. In the afternoon of the day when the *Gazela* came over for a gam, the barquentine's dories and our own were coming back at the same time. Our afterguard stood in a gloomy row by the rail, glowering and muttering to one another about the fullness of the *Gazela's* dories and the emptiness of those from the *Argus*. In fact, though some of the barquentine's were well-filled, a much higher proportion of *Argus* dorymen had done better, as they generally did. Our pessimists pretended not to see it, even when some dories which they had been praising as *Gazela's* when they were a long way off, turned out to be our own. The tune changed and they immediately began to lament their emptiness. No fish! And when young Leite complimented the *Argus* on the good rendering of cod-liver oil, they complained that now they had no barrels left to put cheeks and tongues in. But Leite laughed at them, and they laughed too. They were as bad as a group of Suffolk farmers.

The dorymen were neither pessimists nor optimists. They did not dare to be optimistic, and there was no point in pessimism. The nights were longer now and bitterly cold, with the north wind blowing down fine particles of snow which stung like little needles. The working lights were rigged again for the fish-cleaning. The ship was always rolling now, for it was not calm again. Strong winds which did not reach her sent up a sea which did, and the water was for ever foaming and creaming round the fore-deck. All the throaters, liverers, and splitters had small platforms to keep their feet out of the heaviest water, and the platforms were well battened to allow them to maintain a foothold. They did not notice the motion of the ship or the icy wind. Now and again, a mittened hand would reach up to push a sou'wester from an iron-grey head, while a bearded veteran reached with his split lips for a hand-made cigarette, got a light from a comrade, and worked on. About ten o'clock, the cabin-boy brought a kettle of brandy round and dispensed small noggins to all who wanted them. There was no other break in the work. From down below—very close below, for the hold was almost full— came the salters' cry of "Fish! Fish!"; and for'ard, the flames from the pressure boilers roaring in the narrow

END OF THE CAMPAIGN

chimney threw a fierce light on the oilskinned figures and rugged faces, making them appear like demons.

For hour after hour the splitters and the throaters cut at the fish with their square-ended knives, always with sure, quick strokes; the knives held in bloody mittened hands, the mittens made from a light waterproof canvas. They wore woollen gloves inside the mittens. The big vats were brim full of bloody water which slopped from them at each roll of the ship, and the pounds still half-full of dead cod, hours after the cleaning began. Amid the cod a deckboy squatted, removing tongues—one deckboy to each large pound. On the wooden platform between the trolley-lines, the cabin-boy cut pairs of cod cheeks, rapidly, from a pile of horrible great heads. A line of gaffers and trolley-pushers—all young fishermen, not yet proven men with cleaning knives—passed salt down to the salters, using wooden deck buckets which they sent along from hand to hand. The loudspeaker, clamped to the after bulkhead of the galley, blared endlessly, and the wailing of some great-bosomed female sounded like a banshee screaming in the sodden rigging. All round the gulls cried, swooping down on the titbits of pressed livers which were disgorged by the boilers, and letting everything else go. From the galley came the appetising aroma of the midnight soup of sorrow, simmering there, which the deckboys and some of the dorymen would be too tired to eat when their work was done. The gaffers tossed wet cod into the chutes with their sharp-pointed gaffs, and I wondered how they could use such implements with so much apparent carelessness and yet never have an accident. A gaff could rip a man open, and so could a splitter's knife. Yet none ever did.

The anchor light, hung in the fore rigging, shone brightly on the black sea and the yellow masts, and as far aft as the nests of numbered dories, stacked and lashed down there. Halibut and spotted catfish, for the morrow's bait, washed in the scuppers. Emiliano Martins, who was gaffing fish, took a moment off to dance a few steps to the tune of a wailing Algarvian air. Old Antonio Rodrigues worked hunched up against the wind. His workmate, the young de Sousa, asked hopefully, "Finish tomorrow; no?" but he supplied his own answer. The stern riding light shone on the wet triangle of the big trys'l, to which the schooner was riding. Not far away the cluster lights of the *Creoula, Hortense,* and *Antonio Coutinho* showed that those ships, too, were still working. The radio spoke of a fine fair wind driving the little *Gazela* home: she was running down Davis Straits like a bird with her fill of fish.

"Another week will see us gone," Laurencinha said with a grin.

"Aye, to the Grand Banks," said the boatswain, disconsolately.

We had taken five hundred quintals since young Leite had estimated that seven hundred would fill the *Argus*. He had reckoned without the refrigerator. Captain Adolfo was filling the starboard side of that, and it could stow 200 quintals.

The reluctance of a fishing captain to leave the grounds before his ship is as full as she can be, is understandable, when it is realised that a good day's catch of cod is worth approximately a thousand pounds. Two hundred and fifty quintals of expertly salted cod is equivalent to about 200 quintals of dried fish, and this at an average price of five hundred Portuguese escudos

END OF THE CAMPAIGN

per quintal gives a gross return of more than $3,000, to say nothing of the oil, cheeks, tongues, and backbone membranes, all of which have value. In a ship like the *Argus,* with a good capacity and not too many dorymen, the captain stood to earn a considerable sum from a successful season, for he had a share of everything. A 250-quintal day meant more than $100 to him, and he could earn more than the highest paid ship-master in the regular merchant services from most countries in Europe. But every penny was hard-earned. And this was true for the dorymen, the deckboys, and everyone else.

At last we came to the morning of Sunday, September 10th, a grey, cold day like so many others, with a rising wind, snow-squalls, the sea boisterous and threatening and, apparently, no hope of sailing. There were anxious conferences on the radio before any dories were launched, but the barometer was high and steady and the wind was northerly. The dories were launched at six by general agreement among the ships left: the dorymen were happy to go. They always were. But the wind freshened and the sea got up, and by nine the recall had to be hoisted. Snow-squalls were frequent then and the sea was running ugly whitecaps, with deep troughs. The dorymen had to haul in their long-lines after one cast, and back they came scudding before the wind, the Little King in the van with a handful of cod in the bottom of his dory. It never troubled the Little King that his unfailing ability to be first back when any bad weather blew up, advertised the fact that he could not have had a line down.

It was getting on for midday before the best fishermen were in sight, and good men had been coming back all the morning. The last three were Laurencinha, Salvador Martins, and his brother Estrela. João de Oliveira and Francisco Martins were not far in front of them. They came racing across the foam-streaked sea, with the seagulls crying round them, and as they came nearer we could see that their dories were skimming along like planing dinghies, with tiny jibs set gaily to wind'ard, like spinnakers. Now and again they were lost to sight in the snow-squalls, but when it cleared again, there they always were, racing on. Why had these fearless men chosen to sail so far from the ship that windy morning in September with the ship so near to full? They always sailed a long way: mid-summer or early Fall made no difference to them. It was their belief that they would not find good fish near the ship. But why go ten miles?

Each had a fair load but, on the whole, the fishing was poor that morning, perhaps fifty quintals all together. If they had been able to fish all day the take would have been 150 at least, and that would have filled the ship. Now the weather had broken and it would probably stay broken. We had already had more than our share of luck for a Davis Straits September. The fish were all aboard a little after noon and at once the men set to work cleaning. Recovering some of the dories was ticklish work with the ship leaping and falling about at her anchor. The sea was sweeping the dories so high that some of them had difficulty in avoiding being swept bodily aboard. Sea gushed from every washport and scupper-hole to swamp them, and the ship was rolling wildly. But they were all recovered without damage.

As soon as they were secured, Captain Adolfo weighed. Another shift of anchorage, no doubt, but whither? One place was as good as another. We

END OF THE CAMPAIGN

weighed and drifted; and the younger men found it hard to suppress a certain excitement. Could we be sailing after all? Not waiting for the last few wretched quintals? They scarcely ventured to believe it but whispered the idea one to another, not daring to smile. At the midday meal in the saloon Captain Adolfo, with a grin so Mephistophelean that I, for one, did not know whether to believe him or not, announced that he was done with Greenland, and we were off that very day. The boiled cod tasted like the most luscious of steaks and the claret was like nectar. After the meal, our captain said he was going to sail that afternoon, for Lisbon via Ponta Delgada, but he would not rig down the fish-gear immediately for if the weather improved as we approached the Lille Bank or Fyllas, he would anchor again to take a few more quintals—only a day or so. The men at the fish-cleaning were still saving all the pieces of cods' offal which would appeal to other cod, and no one had taken the bait from the hooks on the long-lines.

Were we sailing, or were we not? We continued to drift, rolling abominably. Then I saw the schooner *Hortense* coming over, and the *Creoula*. The *Hortense* was not yet full enough to think of sailing. O Anjo anchored near-by and sent over two dories for some flour, olive-oil, sugar, and one or two other things he needed. The two sons of Antonio Rodrigues were in the dories. They were fine young men and there was a light in the tired, deep-sunken eyes of their old father as he looked at them. They were good dorymen, both of them.

When the *Hortense* was stored we got under way, setting sail and using the motor, and headed towards the south. So did the *Creoula*. There was a purplish light over the coast of Greenland that late afternoon and the mountains were beautiful; but the only really good thing about that landscape was that we were leaving it, seeing it from the sea afar off, for the last time that season. The northerly wind blew fresh, and for once all hands, except Captain Adolfo, prayed that it would remain so or even freshen still more. It was bitterly cold as the *Argus* and her sister stormed on towards the south, but the dorymen smiled as they steered.

All night we bounded along, and the stars shone. Blow, north wind! Blow, till Fyllas is far behind us! The great white schooners lurched and staggered on, almost scupper-deep with salted cod, without that last hundred quintals. Blow along!

CHAPTER SEVENTEEN

THE VOYAGE HOME

> After the whistling winds have spent their spight,
> On the calm'd Sea the wanton Dolphins play.

WE DID not stay on Fyllas or the Lille Bank or anywhere, except once to heave-to for half a day while the center of a hurricane passed ahead of us. The fresh north wind drove us right down Davis Straits: then it dropped, and the *Hortense* and *Antonio Coutinho*—the last ships there—spoke of good weather and good fishing on Store Bank, and Captain Adolfo pretended to be miserable. The fishing gear was all rigged down once we had passed Fyllas, and this work went with a will. Everything was carefully rigged down, cleaned, greased if necessary, and stowed away—first, the plumbing and the lights: then the vats, the working platforms, the trolley runways, and the pounds. When the weather improved and the water was not so icy cold, a start was made on the dories, but it took many days before the last trace of cod was removed from them and from the decks. All the dories were rigged down and scrubbed thoroughly; the long-lines were hung to dry and then rigged down to their components, the hooks and snoods removed, and everything handed in, individually, by each doryman, to be checked by the mate against a list of what had been issued to him.

For a day or two the men, not daring to believe that the fishing was really ended, worked in their oilskin aprons and full fishing gear, and did not remove anything from their dories. But once Fyllas was passed they burst into snatches of song, though the seas ran high and the schooner, under a press of sail, was shipping them green. The big square running sail was set and the *Argus* bounded along. She was hard to steer, but no one minded that. The dorymen were divided into three watches by night, and there were plenty of men. The night watches gathered round the wheel and yarned of the campaign. Laurencinha said that, other years with an eight-line "trawl" * he had done better than he had this year, with fifteen. He had used fifteen sections of line until September, and then twelve. It was a warm season, he said, but the fishing might have been a great deal better. The dorymen thought that the stocks of cod on the Greenland grounds were declining but they agreed that, if only August had been a month of tolerably good weather, they would have filled the ship by the end of it.

They had, indeed, done very well. The final fishing list posted by Captain Adolfo (who always kept something in hand) showed that his estimate of the cargo

* The dorymen sometimes spoke of the long-line as a "trawl" and referred to trawlers proper as "draggers."

THE VOYAGE HOME

was 12,500 quintals,* of which Laurencinha had taken 450. The average catch of all the dorymen, including the green hands and the Little King, was 235 quintals. Even the Little King was credited with 175 quintals, and he was the worst. Nearly all the other men had qualified for the special bonus paid to those who fished more than 200 quintals. Francisco Martins had fished 325, and João de Oliveira had beaten him for the Second Fisher's place by only twenty quintals. The Fourth Fisher was the cheerful Salvador Martins, a slight and merry little man and a magnificent dory-handler, who was five quintals behind the leading Azorean.

Captain Adolfo called the dorymen into the saloon, one by one, and read them his final estimates. This he was required by law to do. He read out the totals from a list, and the mate read from another list the deductions which had to be made from each man's pay—his advances, charges for radio messages (if any), and so forth. There were few deductions. Captain Adolfo endeavoured to look very fierce as he sat at the head of the saloon table, his swarthy face covered, almost to the eyes, by a dark and luxuriant growth of whiskers. His kindly eyes were half-hidden by a pair of horn-rimmed reading glasses which glinted in the bright electric light. All the lights were on, though the sunlight was streaming through both skylights, and the ship's canary was singing away in his cage, happy at the warmth of the sun. Over the captain's head swung a lemon on strings, brought from Portugal and kept to go back to Portugal, for its days of usefulness as a lemon were long past. Captain Adolfo was dressed in warm but sombre brown,

* The cargo discharged at Lisbon in February, 1951, was 12,880 quintals.

with a leather jacket over his dark shirt. The mate, with his hair and beard handsomely brushed, was in the grey battle dress which is standard in the Portuguese merchant service. Most of the dorymen had spruced themselves for the occasion and some were wearing new shirts.

One at a time, they came into the saloon which some of them seemed almost to fill, for there were many unusually big men among them. I liked the fearless way in which they all looked at the afterguard, like the proud and supremely competent men they were. Obviously, these dorymen had preserved the pride of the skilled craftsman, and they displayed the quiet confidence which characterizes those who have successfully surmounted grave perils. There was not a man there—except perhaps the Little King, and he had almost been run down by a trawler—who had not narrowly escaped drowning at least a dozen times that voyage. There was not a man there who had not his little home and his family to support, and none of them was hearing of a fortune. A few seconds' simple calculation, once they knew the catches with which they were credited, was sufficient to show that what they would receive would be sufficient to support them, and not much more. It was certain that none could afford a holiday beyond the brief three days they allowed themselves on first coming home. No matter: to fish beneath the stars in the clear sky of the Algarve or through the balmy Azorean days, until it was time to go to the Banks again, was recompense enough. Considerable dignity attached to the successful Arctic doryman, and the earnings which accrued to him were big sums in the fishing hamlets.

THE VOYAGE HOME

Watching these men, knowing them a little after the long voyage, I began to understand why the Gremio—and behind the Gremio, the Government of Portugal—was pursuing a policy aimed at the preservation of a traditional craft. I saw, too, that the story of this codfishing campaign of 1950 was in fact the story of an industry revived and rejuvenated without the destruction of its human values or of the human spirit. Modern Portugal—the Portugal of Dr. Salazar—in all its work of renovation has aimed consistently at the retention of what is best from the past.

I could understand, too, what Dr. Salazar had meant when speaking of the establishment of the New State * he said, "We are neither seduced nor satisfied by wealth; by the added comfort which technical achievement brings; by the machine which makes man a less important element; by the craze for mechanization; by brute force, immense, colossal, unique though it may be, so long as these things are not touched by the wing of the spirit and brought into the service of a life which becomes increasingly more beautiful, more generous, and more noble. . . . We seek to make our fields more fertile without silencing the happy songs of the girls who labour in them. . . . The spiritual element which is the source, the soul, the very life of our history, keeps us apart from a civilisation which is going back to barbarism. . . . We do our utmost to preserve . . . the simplicity of life, the purity of custom, the gentleness of feeling, the equilibrium in social relations, the familiar atmosphere of Portuguese life, so modest but

* Quoted in F. C. C. Egerton's "Salazar, Rebuilder of Portugal" published by Hodder and Stoughton, London, 1943.

so dignified, and in this way, through the preservation of our traditions, to maintain ourselves in peace."

Aye, aye, to make the fields more fertile without silencing the songs of the happy girls who labour in them, staunch and robust; to keep the white schooners sailing over the distant seas, and two thousand dorymen, noble exponents of great and ancient skills, afloat in their small boats upon the open, wild, and dangerous Arctic ocean, leading the lives of men. Of what use would it be to convert a Laurencinha into a hauler of nets aboard some modern trawler where his abilities would count for nothing? To disperse the captains of Ilhavo with their knowledge acquired in centuries of tradition? To let the grass grow over the Monica yards, while the rivet-hammers noisily constructed a steel motor-ship to drag its nets wherever they could be dragged and further to denude the feeding-beds? To scrap the schooner fleet and drive the dorymen to be deckhands in trawlers? No, no! These things would not be progress.

A great four-engined air-transport roared overhead, west-bound towards Gander. I knew quite well which form of transport had real mobility. Without all the vast and complicated system of maintenance, without a network of dependable airfields, without a horde of tankers bringing oil, and behind all that, the infinitely more vast and complicated social and financial system which breeds such things and imagines that it needs them, the airplane is a heap of wasted metal on the grass. On the other hand the wind always blows, with brief periods of rest, and is reliable enough. There are fish in the sea, and dorymen will find them. The dorymen in the schooner-fleet that year had fished up 20,-

000 tons of cod without harming the feeding-grounds or the stocks of cod.

That day it was announced that the *Cova da Iria* had foundered. The hurricane we had avoided by heaving-to, she could not avoid. First the giant seas wrenched off her rudder, leaving her to wallow helpless in the sea. Deeply laden as she was with her catch of cod, the wallowing was too much for her. She was a wooden ship of little more than 500 net tons. There was a limit to what she could stand, though she would have been all right if her rudder had not been torn off and damaged the hull as it went. Fortunately the *Inacio Cunha*—another small motor-ship, also well-laden—and the schooner *Adelia Maria* were standing-by, and they took off the crew. The *Cova da Iria* carried sixty-two dorymen and had a full crew of some eighty men. Not a life was lost. The *Inacio Cunha* had been in company since sailing from Store Bank, and the *Adelia Maria*, not far away, was called by radio-telephone. The sea was running wild and the conditions were alarming when the *Cova da Iria's* people had to abandon their sinking ship, but not a dory was damaged at the launching though the seas were sweeping clean over her, and she was then quite unmanageable. There was practically a dory for each man.

Some of the old-timers said the *Cova da Iria* would not have foundered had she been schooner-rigged, for she could then have set storm canvas enough to steady her, rudder or no rudder. This was probably true. But she was built in 1944, when masts and canvas and even cordage for running rigging were scarce and expensive.

The hurricane which sank her had already caused

THE QUEST OF THE SCHOONER ARGUS

other damage to the fishing fleet, for it had passed over the Grand Banks and driven many ships into shelter at St. John's. It had caught the *Aviz, Rio Caima,* and *Infante de Sagres* too far from St. John's for them to seek shelter there. The *Aviz* had lost the best part of her dories and suffered other damage which caused her to abandon the fishing and run for Oporto as best she could. The *Rio Caima,* too, suffered damage, and had to put her helm up and run. The *Infante de Sagres* sprang a serious leak and for a time there was some doubt that she would survive at all. But when the worst of the seas subsided her pumps were able to cope with the situation and she, too, then began the long sail home, though she was not fully loaded. Even the big motor-ships had had to shelter. Now they were complaining of too much bad weather and little chance to fish. But when the odd good day came they were doing very well. There was good cod to be caught on the liners' ground though the trawlers were fishing poorly. There was some talk of a new kind of trawl which would float along just above the rocks. If such a trawl could be perfected, it would be the end of the Grand Banks cod and the grounds in Davis Straits. Already, it was said, there had been some Danish experiments with such a trawl.

While we sailed across the North Atlantic the fish which had been stowed in the starboard refrigerator were brought up, resalted, and re-stowed in the hold, where the shaking-down of the cod had made room for them. There was no longer need to work at the cruel pace which was set on the banks. Once the fishing gear was all down and the dories cleaned and shifted to their passage positions—spread out along the deck, to allow a freer movement fore-and-aft—there was a considerable easing of the routine. Men began to think of home, and some were at work making model dories from pieces of cork, for their children.

The last thing that was done before reaching the Azores was to issue new canvas for the dory sails the following season. This was to enable the men to take the canvas home, where their wives would do the seaming. Some casks of cod-cheeks, tongues, and backbone membranes were prepared for home, but these were the perquisites of the afterguard. A large drum of cod-liver oil was made ready as a gift for the Fishermen's Institute at Ponta Delgada. Even after this, César reported that he had 23,000 litres of cod-liver oil of first quality aboard. César kept a chart at the foot of his bunk, placed so that his eyes rested on it last thing at night and first thing in the morning, on which the ship's homeward track was carefully traced. It bore also the track of the schooner *Creoula* (which, being bound for Lisbon direct, had long since disappeared beyond the horizon on a different course) on a previous crossing, when with brave winds she had run home from Store Bank to Lisbon in nineteen days.

The island of Terceira can never have looked more lovely. It showed first as a dark cloud on the horizon; then, as the morning cleared, Graciosa, São Jorge, and the towering summit of Pico could be seen. All day we had sailed with the land in sight and, by nightfall, there was St. Michaels; and Ponta Delgada was only thirty miles away. The weather was glorious and the sea placid, with a fine sailing breeze: the big schooner

THE QUEST OF THE SCHOONER ARGUS

sailed onwards like a swan. Every stitch of her canvas was set and drawing—all the big fishermen's stays'ls, between the masts, and the enormous outer jib. The sky was a glorious blue, with here and there a cloud of the kind which gives a sure presage of continued good weather. Our Azoreans smiled, and hung about the rail all day looking at their islands, as happy as a party of children going on a treat. The sun shone brightly and the air was balmy.

Had we ever been to Greenland? Were there still dorymen hand-lining from a fleet of schooners, fishing the Arctic seas in a world of airplanes and threats of war? The radio, now and again on the rare occasions when it was not on the Bankers' wave-length, had spoken ominously of the affair in Korea. It had spoken of wars and threats of war, of wars, cold wars and warm wars. The schooner, heeling gracefully in the sea to which she seemed to belong as naturally as a flower to an English meadow, began to appear something of an anachronism in such a world as ours. Anachronism or not, the life it offered was surely the kind of life men ought to live.

Everyone had shaved as soon as we sighted the islands, and the decks were crowded with strange, thin-faced men, the Azoreans dressed in their best clothes, ready to land upon the morrow. Captain Adolfo seemed somehow much older, but the dorymen looked magnificently fit and well. The yellow dog Bobby stood all day by the rail aft, on a drum of cod-cheeks, looking excitedly at the land and sniffing as if he could not bring himself to believe in it.

The *Argus* was thirteen days from West Greenland, beyond the Arctic Circle, to the buoys at Ponta Del-

THE VOYAGE HOME

gada. She arrived shortly after dawn but, even at that hour, two boatloads of the dorymen's friends were out to greet them, and the fishing village at the end of town was sending up fireworks in welcome. In the boats were several lovely small children, including the sons of Raul Pereira and those of Francisco Martins. Raul himself, and Francisco and all the rest of the Azoreans, were beside themselves with delight. Baleia—the Whale— was among the first to hurry off, to see his bride (as well he might), his immersion in the icy Greenland seas now forgotten. The twenty-six Azorean dorymen and the three deckboys from the islands stumbled about the decks, carrying their bedding, clothing, and dory sails to the rail, their faces covered with grins of wild delight while they answered the shouted greetings of relatives and friends, who came alongside in boats. The disembarkation did not take long: we had no other business in the pretty port. But it was evening when we sailed, for the dorymen from the mainland were allowed a brief run ashore. The night was beautiful with the moon almost at full, and there was just enough motion in the sea to set the blocks to creaking, and enough wind to give rounded loveliness to the sails.

Senhor Vasco Bensaude, the principal owner of the *Argus*, was at Ponta Delgada. With him as guide along the roads, past scenery of surpassing loveliness and sometimes down avenues drenched with the perfume of roadside flowers, there was time to have a quick look at the Lakes of the Seven Cities and at famed Furnas, with its bubbling springs and boiling waters, where from clay and rocks a few feet apart there spring forth waters totally different in temperature, taste, and properties—some sweet, some cold; some boiling, some sulphurous; some renowned as a cure for rheumatism, others of value as tonics, or for the skin. Furnas is a strange and lovely place, with its lake of soft water in which, after a few baths, all skins acquire a velvety loveliness. And, as we went there, the views of the island from the winding hill-side roads were beautiful.

But Senhor Bensaude and I spoke much of ships, of schooners and dorymen, and the fishing in the Arctic seas. The Bensaudes had backed ventures in strange ships for more than a hundred years, sending sailing-ships to the whale fisheries, and then Banking. Senhor Bensaude declared that he was more than satisfied with the success of the *Argus* and *Creoula*, and announced his intention to build more schooners of the same type.

With all the Azoreans gone, the ship seemed almost empty of people that night. As we moved out quietly from the harbour, we passed two lateen-rigged fishermen coming in, with sweeps, unlighted. The moonlight touched their sails to a silvered loveliness and, beyond them, fell upon the lovely island; the whole setting was romantic beyond words.

In another four days we were off Lisbon. The *Argus* had blossomed into a mail-carrying vessel at the Azores and had the post aboard: even without this, she would have hurried. The weather remained good. It blew a fine breeze and the big jib blew out; everything else continued to be set and to draw well. The curve of water at the *Argus's* bow sang a merry song, and the sound of the wind in the rigging was a lullaby. And then, at last, the land. The *real* land—home! And the *Argus* was again in the shipping lanes, with the big steamers wallowing along with their holds full of cargo,

THE VOYAGE HOME

and pitching and rolling though the sea was as quiet as a river. The coast of Portugal was a blaze of lights. By nine o'clock on that lovely evening, the last of September, the pilot was aboard and the sails were stowed. There was not a murmur on the Tagus Bar as we came in from sea. The Tower of Belem and the great Church of the Jeronimos were beautiful beneath the moon, and the Tagus was a gentle silvered stream to bear us to our berth. The ship was home.

In the morning, at break of day, a huge tug took the *Argus* from her overnight anchorage off the city of Lisbon and towed her like a big white yacht to the company's moorings in an arm of the river, by Barreiro, whence the trains leave for the Algarve. Here she was secured with chains at a berth astern of the *Creoula* and *Gazela* which had arrived before her; and the ship in her chains should be secure against anything. As the sun shone through the morning mists the sails were loosed for airing, to be unbent later and stowed below. A pretty *fregatta*—a Lisbon sailing barge—came alongside, to carry the men's gear to the railroad station. Accountants arrived from the Lisbon office to pay them off; and the astonishing collection of the dorymen's belongings was carried up on deck, to facilitate the customs inspection. Senhor Gordo Cardoza had been up since four o'clock, frying cods' cheeks and cod steaks from the refrigerator. The men from the shore enjoyed the rare treat of fresh Arctic cod, and several of them ate so many fried cods' cheeks that they had to curl up in sails afterwards, to sleep off the effects. The customs men were courteous and reasonable, and they had all day. The train to the Algarve would not leave before nightfall; neither would the train to Ilhavo, which departed from the station across the river, in Lisbon itself.

It was early evening when I left. The three white sailing ships—*Hortense* was on her way but would not arrive for another week at least—looked beautiful, and strong. They would be rigged down later for the winter months, but for the time being each carried all her masts and yards. Scarred by their six months at sea, rubbed by the ice and a hundred dory launchings and recoveries, rust-streaked a little here and there, they still retained their yacht-like appearance and the air of supreme seaworthiness which had first attracted me to them. The red dories were stowed in their nests aboard —by the mercy of God all the dories with which they had sailed, though the *Creoula* had lost a second doryman in the aftermath of a hurricane on her homeward passage. He was washed overboard. Brave ships! They were well-handled and well-manned and they had been well conceived, too, by a fine maritime Nation, rich in the traditions of great voyages, which refused to be disrupted by the craze for mechanisation and the trend towards barbarism, which elsewhere had driven such lovely ships from the sea.

". . . The simplicity of life, the purity of custom, the gentleness of feeling, the equilibrium in social relations, the familiar atmosphere of Portuguese life, so modest but so dignified . . ." Here were these things all symbolised: here were their fruits in the shape of good ships, good captains, good men, and an abundant harvest brought in from the sea.

The dorymen lined the rail, waving cheerfully. Aft, Captain Adolfo and his mates, all pose of toughness

THE VOYAGE HOME

dropped, were smiling in the sun. The night trains to the Algarve and to Ilhavo would ring merrily that night. At Barreiro the little train was already standing at its platform, and the green, severe compartments looked hard homes for a night—though not, perhaps, to men just landed from an Arctic fisherman, going home after a six-months voyage.

APPENDIX A

THE ARGUS AND HER VOYAGES

The *Argus* is a steel schooner of modern design, built by De Haan and Oerleman at Heusden in Holland in 1938, completed in time to begin fishing during the 1939 season. She was designed by the late Alexander Slater, at the request of Senhor Vasco Bensaude, to replace a predecessor of the same name. This earlier *Argus* was built as a two-masted tops'l schooner (later converted in Portugal to a three-masted plain fore-and-after) at Dundee in 1873. This vessel also took part in the 1950 campaign, under the name of *Ana Maria*, of Oporto: she is one of the last two Portuguese fishing schooners without power, the other being the ex-Newfoundlander *General Rawlinson*, now known as the *Paços de Brandão*.

The new *Argus* was designed to carry about 900 or 950 tons of cod, in salt bulk. Her gross tonnage is 696, and net 414. Her overall length is 209 feet, loaded waterline 173 feet 6 inches, maximum beam 32 feet 6 inches. The steel hull is especially ice-strengthened for working in Arctic seas. The lower masts are steel (the fore lower acts as exhaust for the steam-heating system in the 72-bunk forecastle: the jigger exhausts the main engine, a Sulzer diesel of 475 hp) and the topmasts are of wood. In conformity with Portuguese practice, the after mast is highest, being 119 feet above the deck.

THE ARGUS AND HER VOYAGES

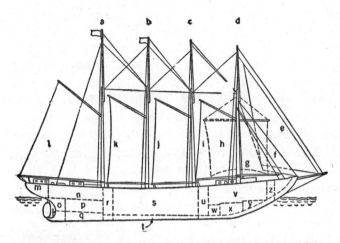

General Arrangement of the Four-masted Schooner *Argus*

a Jigger mast
b Mizzen mast
c Main mast
d Foremast
e Jib
f Inner Jib
g Fore staysail
h Square sail (for running)
i Fore sail
j Mainsail
k Mizzen
l Spanker (trysail set here on Banks)
m Bos'n's store
n Accommodation
o After peak
p Motor room
q Lubricating oil
r Diesel oil
s Fish room
t Fresh water
u Bait (refrigerator)
v Crew space (rancho)
w Coal
x Stores
y Cod-liver oil
z Chain locker

Her sail area is ample, though her sails include no ordinary gaff tops'ls of the type usually associated with big fore-and-afters. Instead, she sets big fishermen's stays'ls between the topmasts, and these have two advantages —they are larger than gaff tops'ls, and they are set and taken in from the deck.

She has a strong power-driven windlass and patent anchors. She has fourteen shackles of cable on the port

THE QUEST OF THE SCHOONER ARGUS

anchor, and twelve on the starboard. She has fuel-oil capacity for 70 tons, fresh-water capacity for 130 tons, and carries 40 tons of bait in a special refrigerated baitroom on the fore part of the timber-lined fish hold. She has two generators, two refrigerator motors, and an electric pump. She is fitted with electric light throughout, with echo-sounding, radio telephony, and separate steam-heating systems for'ard and aft. Her cod liver oil plant of boiler and two pressure cookers produce between 15 and 20 tons of high grade cod liver oil a season. She is flush-decked, with a spoon bow and semi-cruiser stern. Her spike bowsprit is of wood.

The *Argus* belongs to the Parceria Geral de Pescarias, a Lisbon firm which has been sending schooners to the Banks since 1891. Its fleet now includes the *Argus*, her sister-ship *Creoula,* a beautiful three-masted schooner named *Hortense* which was built in Portugal in 1929, and the barquentine *Gazela,* which is the last commercial square-rigger operating out of Western Europe. This firm owns no motor-ships.

The following is a resumé of the *Argus's* twelve voyages:

VOYAGE ONE, 1939

Sailed from Lisbon with full capacity bait in the refrigerator, May 27. Arrived Ponta Delgada May 31 and sailed the same night. Arrived on Grand Banks June 8 and fished there until June 22. Then sailed for Greenland.

Fished in Davis Straits (mainly on Store Hellefiske Bank), June 29 to August 28. Thence returned to Grand Banks. Fished Grand Banks a second time September

THE ARGUS AND HER VOYAGES

4 to October 11, on which day she sailed with a full cargo for Lisbon, reaching Ponta Delgada October 18 and Lisbon October 23, after an absence of a little less than five months and one week.

VOYAGE TWO, 1940

Sailed from Lisbon April 24: arrived Ponta Delgada April 29 and sailed again May 1. Reached Grand Banks May 8 and fished there until May 27. Fished Middle Ground until June 2. Then went into Halifax for bait for Greenland, as the refrigerated compartment could not carry sufficient bait for both the Grand Banks and the Davis Straits fishing, with 53 dorymen each using between 600 and 1,000 hooks, three and four times a day. At Halifax June 4; sailed again June 8; fished another week on Grand Banks and then sailed to Greenland. Fished Davis Straits from June 30 to August 31, then full. Reached Ponta Delgada September 12; returned Lisbon September 18, after an absence of four months and 25 days.

VOYAGE THREE, 1941

Sailed from Lisbon April 23: at Ponta Delgada April 28 and sailed two days later. Fished on Grand Banks May 9 to June 15. Saved some of crew of the Portuguese Banks schooner *Santa Quiteria,* and also a lifeboat of survivors from a torpedoed British steamer. At North Sydney for bait, June 16 to 18. Fished Davis Straits, June 30 to August 15. This was an exceptionally good year of settled weather and abundant cod: by mid-August, the *Argus* had not stowage for another cod. She sailed the night of August 15, reached Ponta Delgada August 31, sailed the next day, and was back at Lisbon by September 5, after an absence of four and a half months.

VOYAGE FOUR, 1942

Sailed from Lisbon May 21: owing to war conditions, did not call at Ponta Delgada. (An entry in the log reads "Boarded at sea by H M Ship, signed F.M. Phillipps, Lt. RNVR," to make sure, probably, that she was not another *Seeadler* with Count Felix von Luckner aboard.) Lt. Phillipps must have been satisfied, for she was on the Grand Banks by June 6. This season she wasted little time there, and touched at North Sydney for bait on June 9. This she must have regretted, for she was delayed there until June 23, no bait being available. Fished Davis Straits from July 2 until September 3, by which date she was full. (There were no trawlers in Davis Straits during the war, and the only other fishermen were Eskimos. Cod were plentiful.) Sailed from Store Hellefiske Bank September 3; reached Ponta Delgada September 17 and Lisbon September 22, after an absence of four months and one day.

VOYAGE FIVE, 1943

Sailed from Lisbon May 31, in convoy. Convoy system had been insisted upon to permit the continuance of the Portuguese codfishing voyages, all belligerents being notified in advance of convoy details. The big schooners and motorships went directly to the Davis Straits grounds, and the smaller schooners to the Grand

Banks. There were no diversions at ports. Reached Davis Straits June 23; fished there until September 8, when full: returned in convoy to Lisbon direct, arriving there October 4, after a voyage of four months and four days.

VOYAGE SIX, 1944

Sailed from Lisbon in convoy May 17. Reached Davis Straits grounds June 8, and fished there until September 9: reached Lisbon on return (full ship) September 30, after absence of four months and 14 days.

VOYAGE SEVEN, 1945

Sailed from Lisbon in convoy—this was the last of the codfishing convoys—April 30. Arrived Grand Banks May 15 and fished there until June 13. Thence to Greenland direct. Fished Davis Straits June 23 to August 24, when full. Back at Lisbon September 12, after voyage of four months and 13 days. An exceptionally good cargo.

VOYAGE EIGHT, 1946

Sailed from Lisbon May 3 and from Ponta Delgada May 9. Fished Grand Banks from May 17 to June 1. At North Sydney for bait, June 3 to June 15. Fished Davis Straits, June 24 to August 27; failed to fill, and so returned to Grand Banks. Fished Grand Banks second time September 7 to September 27. Reached Ponta Delgada October 6, Lisbon October 12, after voyage of five months and ten days.

THE QUEST OF THE SCHOONER ARGUS

VOYAGE NINE, 1947

Sailed from Lisbon May 1, from Ponta Delgada May 9. Fished Grand Banks May 18 to June 4. At North Sydney for bait, June 8 to June 14. Fished Davis Straits June 24 to September 4, when full. Arrived Ponta Delgada September 18, Lisbon September 24, after voyage of four months and 24 days.

VOYAGE TEN, 1948

Sailed from Lisbon April 21, from Ponta Delgada April 26: fished Grand Banks from May 3 until June 4. At North Sydney for bait, June 7 to June 20 (bait late). Fished Davis Straits from June 29 to September 4, when full. Reached Ponta Delgada September 17, Lisbon September 23, after voyage of five months and two days.

VOYAGE ELEVEN, 1949

Sailed from Lisbon April 9, and from Ponta Delgada April 14. Fished Grand Banks April 23 to June 3. At North Sydney June 4 to June 18 (bait again late). Fished Davis Straits June 28 to September 15 (much bad weather, and fish not plentiful as during the war years). Returned to Grand Banks, and fished there again from September 25 to October 7. Reached Ponta Delgada October 13, Lisbon October 18, after voyage of six months and nine days. Not quite a full cargo.

VOYAGE TWELVE, 1950

Sailed from Lisbon April 1, and from Ponta Delgada April 8, bound for St. John's, Newfoundland, for bait.

THE ARGUS AND HER VOYAGES

Arrived St. John's April 15; delayed there: sailed again May 3. Fished Grand Banks May 4 to May 31. Thence to North Sydney, June 3 to June 12. Fished Davis Straits June 21 until September 10: Sailed thence for Ponta Delgada, where arrived September 23. At Lisbon September 30, after voyage of six months. Full cargo.

APPENDIX B

THE PORTUGUESE GRAND BANKS FLEET

IN 1950, Portugal sent a fleet of sixty-two vessels to fish for cod on the Grand Banks and off the west coast of Greenland. These included eighteen trawlers of some 1,200 to 1,500 tons, of modern design; thirteen motor-ships especially designed for fishing from dories; and thirty-two sailing vessels, of which two were without auxiliary power. This compares with a fleet of fifty-seven vessels in 1936, none of them trawlers, none motor-ships, and only twenty-two of them with auxiliary power. The thirty-two sailing vessels in the 1950 fleet include fourteen four-masted schooners and one barquentine. The remainder were three-masted schooners. The 1936 fleet included several tops'l schooners, both two and three-masted. This is a rig now discarded, although the present fleet includes at least two vessels which began their careers as two-masted tops'l schooners.

The sailing-vessels in the Portuguese fleet fall into three main classes—old three-masters mainly acquired from abroad; modern steel vessels, rigged as four-masted schooners but virtually full-powered motor-ships as well; and wooden three-and-four-masters built during recent years in Portugal (the best of them by the well-known firm of Monica at Gafanha, near Aveiro), many of most modern design.

The modern wooden fleet, built in Portugal, includes

THE PORTUGUESE GRAND BANKS FLEET

the very handsome pair of Monica schooners, *Brites* and *Novos Mares*. These 420-tonners are as distinctive and as modern as the *Argus* and her sisters. Their high masts and tall topmasts give them ample sail area, which they are accustomed to use, and their flush decks, pleasing sheer, and finely lined ends give them an attractive appearance. Painted white and always well kept, they have the appearance of large yachts. The *Brites* and *Novos Mares* are rigged with the conventional jib-boom, but later Monica vessels are designed to carry their headsails inboard, doing away with the protruding spar which for so many centuries has been one of the features of the sailing-ship. Later ships of this type have discarded fitted topmasts as well; but none-the-less their general appearance remains pleasing enough. Their able, graceful hulls combine with the utilitarian but still symmetrical and ample rigging to give them good profiles.

One of the best ships of this type is the *Dom Deniz*, of 530 tons, built by Monicas at Gafanha in 1940.

A list of the 1950 sailing fleet follows:

THE QUEST OF THE SCHOONER ARGUS

PORTUGUESE GRAND BANKS FLEET

SHIP	TONNAGE	RIG
ARGUS	696	Steel four-masted schooner
CREOULA	665	" " " "
SANTA MARIA MANUELA	666	" " " "
JOSÉ ALBERTO (ex-CAROLINE)	687	" " " "
MILENA (ex-BURKELAND)	757	Wood " " "
BRITES	423	" " " "
NOVOS MARES	434	" " " "
GROENLANDIA	442	" " " " (built as 3/m tops'l sch.)
SENHORA DA SAUDE (ex-HELGA)	427	Wood four-masted schooner
AVIZ	523	" " " "
CONDESTAVEL	635	" " " "
ADELIA MARIA	667	" " " "

THE PORTUGUESE GRAND BANKS FLEET

BUILT	NUMBER OF DORYMEN CARRIED	NOTES
Holland, 1938	53	Took full cargo, over 10,000 quintals in Davis Straits.
Lisbon, 1937	52	Full cargo, 9,000 Q. from Davis Straits.
Lisbon, 1937	57	Almost a full cargo.
Marstal, 1922	54	Fished on Grand Banks twice: not full cargo.
Milton, Florida, 1918	58	Fished on Grand Banks twice.
Gafanha, Portugal, 1936	46	Fished on Grand Banks twice, after a late start.
Gafanha, 1938	45	Sailed from Davis Straits September 15, the last to leave this area.
Vila do Conde, 1922	45	Fished on Grand Banks twice.
Fanö, Denmark, 1920	48	A good season: filled in Davis Straits.
Gafanha, 1939	51	Returned to Grand Banks a second time: lost dories in hurricane and had to abandon fishing.
Gafanha, 1948	56	One of first ships to complete, in Davis Straits.
Gafanha, 1948	52	Did well: completed cargo off Greenland.

THE QUEST OF THE SCHOONER ARGUS

PORTUGUESE GRAND BANKS FLEET (continued)

SHIP	TONNAGE	RIG
COIMBRA	668	" " " "
VIRIATO	594	" " " "
GAZELA PRIMEIRO	320	Wood three-masted barquentine
ANA MARIA (ex-ARGUS)	271	Wood three-masted schooner
ANA PRIMEIRO (ex-ERIKA)	275	" " " "
ANTONIO RIBAU	366	" " " "
CRUZ DE MALTA	296	" " " "
DOM DENIZ	530	" " " "
HORTENSE	874	" " " "
INFANTE DE SAGRES	330	" " " "
LABRADOR (ex-LYDIA)	307	" " " "
LOUSADO (ex-ALCION)	224	" " " "

THE PORTUGUESE GRAND BANKS FLEET

BUILT	NUMBER OF DORYMEN CARRIED	NOTES
Gafanha, 1948	57	Did well: completed cargo off Greenland.
Gafanha, 1945	52	Returned a second time to Grand Banks.
Cacilhas, 1883: rebuilt at Setubal, 1900	30	Full cargo from Greenland.
Dundee, 1873	28	Did not go to Greenland.
Sweden, 1918	31	Burnt out and lost on Grand Banks, August, when practically full.
Gafanha, 1921	40	Fished full off Greenland: one of first ships to leave.
Gafanha, 1921	34	Fished full on Grand Banks: did not go to Greenland.
Gafanha, 1940	47	Fished full off Greenland.
Gafanha, 1929	33	One of last to leave Greenland.
Gafanha, 1921	32	Fished twice on Grand Banks, but driven off with weather damage: returned Portugal.
Korsor, Denmark, 1919	32	Returned Grand Banks: did not fish full. This vessel was using long-lines for first time.
Cumberland, 1904	28	First ship to fill, and first ship home. Fished Grand Banks only.

THE QUEST OF THE SCHOONER ARGUS

PORTUGUESE GRAND BANKS FLEET (continued)

SHIP	TONNAGE	RIG			
MARIA DAS FLORES	607	"	"	"	"
MARIA FREDERICO	469	"	"	"	"
OLIVEIRENSE	421	"	"	"	"
PAÇOS DE BRANDÃO (ex-GENERAL RAWLINSON)	187	"	"	"	"
RIO CAIMA (ex-SANTA MAFALDA)	353	"	"	"	"
RIO LIMA	368	"	"	"	"
SAN JACINTO	248	"	"	"	"
SANTA ISABEL	345	"	"	"	"

In addition to these thirty-two vessels, five other three-masted schooners, all former Grand Bankers, were in existence, but did not take part in the 1950 campaign. These were the JULIA I, TROMBETAS, NEPTUNO II, ILHAVENSE II, and NAVEGANTE, which were laid-up as unserviceable. The NEPTUNO II was written-off as a constructive total loss.

THE PORTUGUESE GRAND BANKS FLEET

BUILT	NUMBER OF DORYMEN CARRIED	NOTES
Murtosa, Portugal, 1946	45	Fished full off Greenland.
Gafanha, 1944	47	Fished full on Grand Banks, and early home.
Gafanha, 1938	44	Fished full off Greenland.
Newfoundland, 1920 (rebuilt Villa Nova de Gaia, 1923)	24	Fished full on Grand Banks: a good season.
Gafanha, 1929	39	To Grand Banks twice; lost dories second time and had to return to Portugal.
Viana do Castelo, 1920	37	First ship to sail from Greenland.
Gafanha, 1945	32	Fished full on Grand Banks: did not go to Greenland.
Gafanha, 1929	35	Fished full off Greenland: a good season.

APPENDIX C

SOME ECONOMIC NOTES

EVERY man in the fishing fleets is registered at the central bureau of the Gremio—the Guild of the Codfish Shipowners—in Lisbon, and has a number. In theory, dorymen are directed to their ships, and crew lists are furnished for each campaign from the Guild, but in fact it is the general rule that crews change little, except through natural wastage. In the *Argus*, the majority of the men had made every voyage since the ship was built, and the Gremio would not think of directing them to any other vessel. Each man's catch is recorded, and he is graded as Special, First, Second or Third Line. He takes the grading of his previous campaign when he signs for the following year. The grades are based on the number of quintals of green—or salt bulk—cod the man caught, and they are as follows:

Special	Those	who	took	over	153	quintals
First Line	"	"	"	from	115 to 153	quintals
Second Line	"	"	"	"	85 to 115	quintals
Third Line	"	"	"	"	60 to 85	quintals

Any who caught less than sixty quintals would not be fishermen at all. The only other grade was that of Green Fishermen, which covered first voyagers. The *Argus* had only Specials, and Green Hands. The qualifying catches were based, I believe, on the older methods of hand-lining, and a Greenland fisherman, using at least 600-hook lines, ought to take more fish, as in fact he does.

The men are paid in accordance with the terms of an agreed contract which is made between their representatives and the Guild, and printed. A copy of this contract is in all the ships, and the men have access to it. The voyage articles which they sign refer to their remuneration as "according to contract," and this is accepted. All engagements—cooks, engineers, deckboys, officers, and dorymen—are regulated through the Guild, and there is no floating force of casual labour such as is general in other branches of seafaring. In addition to his pay, each doryman is entitled for each campaign, to the following:

- 1 pair of thigh sea-boots (rubber)
- Canvas to sew new sails for his dory
- 1 sou'wester
- 7 lbs of tobacco
- 8 litres of oil, for treating oilskins, sails, and fishing aprons
- 6 pairs of gloves

His pay is made up from sums received for his skill as splitter, salter, or whatever he may be; from an advance, before joining; and from sliding-scale payments on his catch, at so much a quintal as estimated by the captain at the time the fish is taken. Every doryman is a fish-cleaner or salter of some sort, in addition to being a fisherman, and some are also sailors. The same man may be a sailor, a splitter first class, and a doryman special class. He receives pay for each qualification. Splitters and salters are graded in classes according to their experience and speed of working: the higher the grade the better the pay, though the difference is not great. According to his classification and grade, each man receives a fixed sum for this part of his services for the whole campaign. These payments (given in Portuguese escudos, which in 1950 were worth 80 to the English £ and about 27 to the American $) are as follows:

CLASSIFICATION	FIXED SALARY, PAID AS AN ADVANCE, BEFORE SAILING	NOTES
Boatswain	5,900$00	Also paid as a doryman and as splitter, first class.
Cook	5,500$00	Not being a doryman, is paid a share of the total catch: this share being at the rate of 1$ 20 for each quintal of cod discharged at Lisbon.
Assistant Cook	4,200$00	His share is based on the rate of 70 Centavos a quintal of cod, as landed at Lisbon.
Assistant Engineer	Same advance and share as Assistant Cook.	
Deckboys	3,550$00	At the end of the voyage deckboys receive a further lump-sum of 2000$00 to 2500$00 according to their diligence and efficiency.

SOME ECONOMIC NOTES

CLASSIFICATION	FIXED SALARY, PAID AS AN ADVANCE, BEFORE SAILING	NOTES
Sailors	5,000$00	Plus lump sum of 250$00 for his special duties in this category: plus earnings as dorymen.
Dorymen	5,000$00	Plus earnings for catch, and lump sum as salter, splitter, etc.
Green Dorymen	4,200$00	Plus catch.

In addition, salters first class receive a lump sum of 500$00 for the campaign, salters second class 450$00, and salters third class receive 350$00. Splitters first class receive 400$00, splitters second class 350$00, and splitters third class (of whom there were none in the *Argus*), 250$00. Throaters receive extra payments of 150$00 only, while gaffers, trolley-pushers, passers of salt, and all others engaged purely in labouring duties during the fish-cleaning, receive extra payments of 50$00.

Payments to dorymen, and all employed or acting as dorymen, are as follows:

```
For each quintal of green cod, landed, up to 100 Q.    20$00
  "    "    "     "    "    "    "  , from 100 to
                                              150 Q.    26$50
  "    "    "     "    "    "    "  , from 151 to
                                              200 Q.    32$00
  "    "    "     "    "    "    "  , above 200         35$00
Additional Bonus to each man taking 200 Q. or more     200$00
```

The payment for the catch is subject to a sliding scale on a cost-of-living basis, and during the 1949 and 1950 campaigns these rates were increased by 65%.

To give an idea of the actual returns to the men, the First Fisher of the *Argus* received, for the 1949 season, payment as follows:

Payment (fixed) as Fisher, on signing		5000$00
" " " Sailor		250$00
" " " Splitter, First Class		400$00
Catch		
100 Q. at 20$00	2000	
50 Q. at 26$50	1325	
50 Q. at 32$00	1600	
256 Q. at 35$00	8860	
	13785 Plus 65%	22845$00
	BONUS	200$00
	Total	28695$00

From this total of 28,695 escudos, about 1000$00 is deducted for various charges and taxes, for the Fishermen's Institute insurance, and so on. All the taxes are in his own interests. His net pay that season was therefore about 27500$00—approximately £345 in English money, or $970 in United States dollars. He could not have earned anything like this sum in any other way. He received, in addition, full subsistence during the voyage, and medical care for himself through the *Gil Eanes* and the team of Guild doctors, and for his family, through the Fishermen's Institute in their home town. He could avail himself of special housing facilities, schools for his children, and various other benefits, all arranged by the Fishermen's Institute. He would receive a small pension at the age of 64. His place in the *Argus* was therefore worth considerably more to him than the sum of actual cash he received. As a measure of the economic position he was able to attain through his work as a fisherman—both from Fuzeta beach and on the Banks—it may be mentioned that this man, beginning with nothing, owned three houses at the age of 35. His 27,500 Portuguese escudos were worth far more to him, and to his family, living quietly at Fuzeta, than $970 would be to any American family, or £345 to any English. His cost of living was much lower, and his money would go much further.

The First Fisher of the *Argus* was an exceptional man. The average cash return to dorymen was considerably less. A man who took 200 quintals would receive some 13000$00, or about $480. But the same remarks apply to him, and he enjoyed the same benefits.

Since the total catch is never known until the cargo is discharged, and that might not be until some months after the ship's return to her home port, it is customary to pay the men on the Captain's estimate, and then to adjust their pay when the precise quantity of fish is known. The adjustment is made in accordance with the man's catch as estimated. All the captains try to keep some percentage in hand. It is impossible to estimate precisely, and to adopt any form of measuring on the Banks would slow the work to an unacceptable degree. The present method works quite well, and the men receive the extra money at a time when their families have greatest need of it—during the winter.

Captains and officers are paid on shares of the ship's total cargo, and so also are cooks. In low-powered schooners, the engineer is paid the same share as the chief cook but, in ships with greater horse-power, his share is more. In the *Argus*, for example, with an engine of 475 hp, the engineer in charge received a total return for a voyage yielding 12,000 quintals of fish—a good cargo: but she usually discharged more than this, and sometimes considerably more—equal to about £500 English, or $1400 in U. S. dollars at the then rate of exchange. This was for his whole year's work, because he is permanently employed. His fixed salary (for the campaign) was 6250$00 to which was added some 35000$00 as his share of the catch, including the cod-liver oil. Taxation has been allowed for in the estimate of £500 as his return.

Captains and deck officers are well paid, particularly captains. The officers receive fixed advances for the voyage, but the captain has a monthly salary, payable throughout the year, in addition to his share of the catch of cod, the take of cod-liver oil, and the salted tongues, cheeks, and *samos*. In a ship which took 12000 quintals, with about fifty dorymen, the captain's gross return would be at least £2000, which makes him one of the best-paid master mariners at sea today, at any rate sailing out of Europe. He is paid on the basis of an agreement negotiated between the National Syndicate of Captains and Officers, and the Guild, and his share is subject to a sliding scale based on the cost-of-living, in the same manner as the men's. His fixed salary is in the region of 1200$00 a month, and his share of the cod—based on a thirty-doryman ship—is 9$00 a quintal for the first two-thirds of the cargo landed and 13$00 a quintal for the rest. There was a 50% advance on this in 1950. He receives fifty centavos

SOME ECONOMIC NOTES

a kilogram for the cod-liver oil, and a third the value of the other products.

This share on the cod is a standard share, and adjusted according to the number of dorymen carried. For instance, the actual share in the *Argus* was thirty fifty-thirds of standard. In a seventy-doryman ship, the captain's and officers' shares would be thirty-seventieths of standard. To offset this, of course, they would share in a much larger cargo, and the purpose of the standard share is to bring about a rough equalisation throughout the fleet. Officers in small ships worked just as hard as in the larger, if not harder, because the ships under 500 tons carry only one mate. Some captains have owning interests, and in that way have a direct stake in the outcome of the voyage, in addition to their own shares.

The standard rates for mates include additional payments, on a monthly basis, while employed in the ships, apart from the time spent on the actual voyage. A mate is usually required to join his ship two or three months before she sails, and a second mate—if carried—perhaps a month earlier. Many of the officers are young and comparatively inexperienced, and the returns from a successful cod-fishing voyage are higher than they could earn in other ways. In the winters, some of them serve as officers in coastal and short-voyage trawlers, or as relieving officers in merchant ships, usually trading to the Portuguese Atlantic islands.

All the captains, and most of the officers, live in excellent homes, the great majority at Ilhavo. If they care, some have beach homes at Costa Nova nearby, and a few have small yachts, and automobiles. Good schools, primary and secondary, are available for their children. By judicious investment in the shares of the ships they run, the way is open for any competent officer to rise, and many owners were former masters. For the doryman, the way of advancement is by accumulating funds to buy a little fishing-boat of his own, and many of the Fuzeta-men and the Azoreans have done this.

Agreements with the officers and the men cover such things as replacement of lost kit if ships are sunk, incapacity pensions, death payments to widows, and so on. It is agreed that there shall be no salvage claims among the fishing fleet: all undertake—and that cheerfully—to render assistance whenever it may be needed to the best of their ability without any question of payment.

The total crew costs—wages, free issues, food and wines, medical stores—in such a ship as the *Argus,* on an average six-month voyage, would not be less than a million and a half of Portuguese escudos—about $60,000. In addition, the owner has to provide for depreciation (a heavy item with new ships), insurance, dues, charges, stores, drydocking, bait purchases, office overheads, outfit of fishing-gear (a heavy item), dory upkeep (a one-man dory which used to cost £6 or £8 a few years ago now costs at least £25), ground tackle (always some is lost), salt by the hundred tons, coal, water, fuel-oil, and the hundred other things which a ship needs. His ship can load once only in the year: when she is not fishing, or on her way to and from the Banks, she is idle and earning nothing. Her replacement costs today are enormous, almost frightening. A good ship which would have cost £30,000 ten years ago now costs so much that it would be quite uneconomic

SOME ECONOMIC NOTES

to build her: an indifferent substitute, with neither the workmanship nor the materials put into her, would cost at least £100,000.

The owner's return can come only from the sale of cod and cod by-products, and the price of these is fixed. He sells his cod as dried *bacalhau*, and a cargo of 12,000 quintals might sell only as 9,000, or even less. The crew are paid for the green fish, and the owner has to allow for the loss by drying. The value of a full cargo in the *Argus*, in the 1950 season, was between £55,000 and £60,000, in English money. Dried cod was sold retail for about one shilling a pound—12$00 to 13$00 a kg.—but the owner's return was at the rate of about 8$00 a kg. This was after he had dried the fish, and he had to maintain a staff and a drying-plant for that, or meet the charges of some other plant.

For his return, he has to risk a considerable investment, to risk his ship in the ice, and to wait a long time for his money. He has to set aside a considerable sum against replacement of his ship (or face heavy interest charges for loans). Fuel and stores costs are steadily advancing. And then, how long may his product be marketable? The taste for salt cod has largely disappeared in several countries already, and better marketing of more abundant supplies of fresh fish may, in time, cause a heavy shrinkage of the markets in Portugal. The dried cod is not particularly cheap, nor is it especially nourishing. If fish must be brought back fresh, the hand-liner might go out of business. An investment sufficient to equip a ship to carry 12,000 quintals of quick-frozen fish would be too great to pay on an annual six-months voyage to collect the cargo

THE QUEST OF THE SCHOONER ARGUS

and there would have to be a faster turn-round of the ships and more voyages.

Yet it is probable that the dried cod trade will last for some time yet, at least among the hardy Portuguese, the Icelanders, and the Norsemen.

THE END

INDEX

Abelhart, 40
Abilio, of the Argus, 274
Adelia Maria, the, 248, 303
Adolfo, Captain, 13, 33, 58, 97, 103, 141, 144, 147, 150, 161, 178-180, 195, 208, 272, 310; knowledge of cod's habits, 117; discusses Canadian schooners, 120; as leader of men, 129, 130; responsibilities of, 130 ff.; commodore of fleet, 139; recounts wartime voyages, 145, 146; discusses Banks fishing, 156, 157; as captain of *Creoula*, 194; career at sea, 217 ff.; knowledge of Store Banks, 284, 285; estimate of fish take, 288, 289, 299
Aguas Santas, the, 141
Algarve, province of, 24
Allen, J. W., 13
Almeida, Captain, 13, 37, 72, 131, 139, 219, 222, 286
Almeida, Commander Americo Angelo Tavares de, 12, 28, 140, 142, 187, 243 ff., 261, 265, 280
Alvaro Martins Homem, the, 242
Ambassador, Portuguese, to United States, 11, 50
America, the, 218
Ana Maria, the, ex-*Argus*, 312
Ana Primeiro, the, ex-*Erika*, 267, 268
Anarhichas Minor, see Catfish

Andes, the, 20
Anibal, Captain, 193-195, 197, 221
Antonio Coutinho, the, 286, 292, 297
Antonio Ribau, the, 171, 280
Arctic Circle, the, 155, 171, 181, 184, 185
Argus, the, 11, 16, 168, 170, 238, 252, 272, 289; described, 19 ff., 49, 50, 312; arrival in Azores, 51; equipment of, 54 ff.; life on, 65 ff., 111, 112; senior ship in fleet, 139; strengthened against ice, 161; dorymen of, 197; Captain Adolfo becomes master of, 221; captained by Captain Vitorino, 244; capacity of, 278, 293; returns to Lisbon, 309; the earlier, 312; diagram of, 313; resumé of Banks voyages of, 314 ff.
Arruda, Roberto, 13
Aveiro, Portugal, 11, 24
Aviz, the, 72, 139, 147, 160, 161, 164, 286, 304
Azores, First Fisher of, see Martins, Francisco
Azores, fishermen of, 52

Bacalhau, 44 n.
Baffin-land, 154
Baffins Bay, 158
Bait, 69 ff., 85, 124 ff., 152, 176

INDEX

Baleia, called the Whale, 232-234, 307
Banker, French, 235
Banquereau, the, 144, 192
Barents Sea, 123, 266
Barges, Lisbon sailing, 22, 309
Barquentine, 206; *see also Gazela, Terra Nova* (old)
Barreiro, Portugal, 309, 311
Basques, early, 43
Battista, family of, 170
Battista, Francisco Emilio, First Fisher of *Argus, see* Laurencinha
Battista, José (the elder), 189
Battista, José (the younger), 193
Battista, Leandro, 189
Battista, Manuel, 189
Belem, Portugal, 19, 30
Belem, Tower of, 35, 309
Belle Isle, Straits of, 153, 154, 158, 162, 163
Benac, M., 13
Bensaude, Dr. Joaquim, 12
Bensaude, fleet of, 193, 212, 221, 286, 288
Bensaude, Vasco, 12, 50, 51, 194, 212, 307, 308, 312
Bensaudes, the, enterprise of, 260
Berlin, the, 146
Berthoud, Pierre, 17, 18
Bianco, Andrea, 45
Biologist, Danish Government, 264
Bismarck, the, 211
Blessing service, 21, 27 ff.
Boat, Eskimo fishing, 237, 238
Brazil, official discovery of, 45
Brazil, Portuguese sailing under flag of, 220

Bremerhaven-Geestemunde, 242
Brites, the, 286, 321
British Sailors' Society, mission of, 151
Buoys, cork, 154, 155
Buques, 190-192, 197

Cabin-boy, of *Argus*, 232
Cabot, John, 43
Cabral, José, 231
Cabral, Pedro Alvares, 45
Caiques, 190, 191
Camoens, Luis Vaz de, author of *Lusiads*, 14
Canada, derivation of name of, 45
Canoa, 191, 197
Cape Bojador, 242
Cape Breton Island, 147
Cape Farewell, 40, 162, 171
Cape Race, 40
Cape Spear, 68
Capitão Ferreira, the, 72-74, 172, 181, 286
Captains, Portuguese Arctic, 149, 150, 151, 250
Caravel, 40
Cardozo, Manuel Gordo, 228, 234, 309
Cargo, final estimate of, 299, 300
Carmona, President, 200
Cascais, Portugal, 36
Catfish, 175, 176, 287
Chalão, Antonio Rodrigues, 256-258
Chief of the Services for Assistance to the Ships at Sea, the, 11, 12
Clara, the, 220
Cocos Islands, 213

INDEX

Cod: cleaning and salting of, 93 ff., 290-292; converting livers of into oil, 96; eating habits of, 117, 118
Cod Conservation Conference, 265
Coimbra, the, 72, 274, 280
Columbus, Christopher, 37, 38, 42, 46
Compass, adjustment of, 22
Condestavel, the, 128, 148, 217, 236, 248, 274, 277, 278
Convoys, sailing, during War, 245, 246, 316
Costs, total crew, 336
Cova da Iria, the, 70, 72, 217, 286, 303
Coward, E., 162, 163
Craft, beach fishing, Portuguese, 150
Creoula, the, 13, 51, 90, 116, 138, 139, 141, 144, 147, 170, 295, 305; sister-ship of *Argus*, 34, 68, 154, 314; loss of dorymen, 132, 310; in fog, 160; in storm, 164; first of modern steel schooners, 194; earlier, 209; commanded by Captain Adolfo, 221; commanded by Captain Vitorino, 244
Cruz de Malta, the, 219
Cunha, Mariano da, 178-180
Cycle, warm, 156, 265, 266

D. Fernando y Gloria, the, 24, 28
Damaso, Francisco de Sousa, 107
Dana's Bank, 155
Davis Straits, 11, 20, 137, 153 ff.; 171, 225, 297; discovery of, 41
Deckboys, duties of, 113, 230-232
De Laes, the, 245, 269
Dias, Bartolomeu, 45
Dighton Rock, 46
Disko Bay, 155
Dogs, of *Argus*, 160, 178, 218, 306
Doldrums, the, 46
Dom Deniz, the, 108, 109, 280, 321
Dories, 16; one-man, 17; described, 21; preparation of, 60, 61; contents of, 87; method of launching, 88; method of hoisting inboard, 93; two-men, 120-122, 203, 235; origin of, 150
Dory, crossing of Atlantic in, 100 n.
Dory-fishing, preparation for, 54 ff.; dangers of, 104 ff.
Dorymen: Azorean, 52; lost at sea, 132; endurance of, 183 ff.; Portuguese Arctic, 188 ff.; dress of, 234

Eanes, Gil, 242
Egerton, Colonel F. C. C., author of *Salazar, Rebuilder of Portugal*, 14, 301 n.
Elisabeth, the, 13, 34, 35, 90, 154, 170, 205, 222, 236, 252; fleet surgeon in, 141; in fog, 160; good fortune of, 177; leaves Banks, 277, 278
Engineer, assistant, 129
Erebus, the, 163
Eric, King of Denmark, 40

341

INDEX

Erika, the, 267
Ernani, the, 269
Estavão Gomes, the, 242

Faeringerhafen, Iceland, 165
Fanshawe, Sir Richard, 14
Faro, Portugal, 191
Fécamp, France, 16
Fernandes, Francisco, 43
Fernandes, João, 43
Fernandes Labrador, the, 242
Figueria da Foz, Portugal, 11, 24, 27, 148, 171, 267
Finisterre, 46
Fish, rolling down of, 227
Fisher, First, of *Argus*, see Laurencinha
Fisheries Production Committee of the Combined Food Board for the United Nations, the, 13
Fishermen, Arctic, clothing of, 22, 23
Fishermen's Institute, at Ponta Delgada, 305
Fishing, long-line: method of from dory, 172 ff.; Algarvian, 194; Canadian, 194
Fleet, 1950 Grand Banks, listing of, 322-327
Fleet, last working sailing, 21
Fog, 100 ff., 144 ff.; dorymen lost in, 251 ff.
Franklin, Sir John, 163
Fregatta, 309
Fructuosa, Gaspar, 44
Furadouro, Portugal, 24
Furnas, Azores, 307, 308
Furness Withy Line, the, 13
Fuzeta, Portugal, 24, 27, 189; fishermen from, 104
Fyllas Bank, 155, 167, 169, 171, 266, 295, 297

Gafanha, Portugal, 26
Gam, 288 n.
Gama, Vasco da, 45
Gamo, the, 209, 221, 244
Garrocho, Manuel Martins, 191
Gaspar, the, 231, 269
Gazela, the, 13, 26, 139, 164, 206, 209, 211, 216, 274, 286, 289, 290, 314
Genepesca I, the, 285
Genepesca IV, the, 81
General Rawlinson, the, 312
Gilbert, Sir Humphrey, 69
Gil Eanes, the, 12, 13, 19, 28, 187, 211, 265-267, 270, 280, 281; arrival on Banks, 140; described, 141, 142; services to the fleet, 241 ff.
Gloucester, Massachusetts, Portuguese fishing from, 214
Godhaven, Greenland, 245
Godthaab fjord, 155
Gonçalves, João, 43
Government Fisheries Laboratory, the, 13
Grace Harwar, the, 18
Graciosa, 305
Graf Spee, the, 209
Grank Bankers, 16
Grand Banks, the, 11, 157, 158
Great Ice Barrier, 162, 172
Greenland: fishing grounds, 11; fishing off coast of, 153 ff.; Banks, 155; Administration, 265
Greenlanders, 264
Gremio dos Armadores de Navios da Pesca do Bacalhau, 11, 28, 43, 193, 221, 328; aims of, 48 ff.; inter-

est in two-men dories, 121;
representative of, 140;
early opposition to, 199;
support of, 212, 301; and
juvenile recruitment, 220;
professional school of, 231;
institutes sick pay, 248
Groenlandia, the, 108, 286
Grosvenor, Melville Bell, 14
Growlers, 158
Guild of the Codfishing Shipowners, the, *see* Gremio dos Armadores de Navios da Pesca do Bacalhau
Guinea, Gulf of, 46
Gushue, R., 13

Haan, de and Oerleman, 312
Hansa Line, 242
Hansen, Dr. Paul, 264, 265, 271
Hansobe, the, 165
Helluland, 39
Henry, Prince, the Navigator, 39 ff., 242
Heusden, Holland, 312
Holsteinsborg, Greenland, 244, 245, 261 ff., 270, 271
Holsteinsborg Bay, 155, 181, 206, 218, 225
Homem, Alvaro Martins, 41, 44
Hora de Saudade, 110, 141
Hortense, the, 139, 162, 164, 166, 193, 194, 197, 236, 286, 295, 297; described, 26; built, 221, 314
Hull, North Sea fishing, 264
Huskies, 270

Icebergs, 158 ff.
Iceland, 123
Ice-shoving, 160, 161

Ilhavo, Portugal, 13; mariners of, 27, 212-214; winter occupations of men from, 104; seafaring traditions of, 149
Inacio Cunha, the, 189, 286, 303
Infante de Sagres, the, 146, 222, 284, 304
Institutes, Fishermen's, 23
Insurance, Fishermen's Institute, 332
Isortok, Greenland, 215, 225, 237, 273, 282

Jeronimos, Church of the, 27, 29, 309
Jig fishing, 70, 248, 274; method of, 239
João Alvares Fagundes, the, 242
João Corte Real, the, 242
João Costa, the, 284
José, nephew of Laurencinha, 189
José Alberto, the, 286
Julia IV, the, 269

Korea, war in, 247, 306

Labrador, 153
Labrador, the, 245, 252, 284
Labrador Current, 154, 159, 161, 163, 171
Labrincha, Captain, 72, 139
La Frontera, Pedro de, 42
Lahneck, the, 242
Lakes of the Seven Cities, the, 307
Lance, 237, 238, 239
Las Casas, 42
Laurencinha, 36, 142, 160, 170, 174, 175, 224, 227,

INDEX

294; sea experience of, 188 ff.; almost drowns, 285; total catch of, 298, 299; earnings of, 332
L'Aventure, the, 28
Lavrador, Fernandes, 45
Leite, Captain José Teiga Gonçalves, 13, 34, 211, 214-217, 286, 288
Leonard Brothers, 148, 152
Lt. René Guillon, the, 72, 235, 250
Lille Hellefisk Bank, 155, 171, 175, 295, 297
Line-fishermen, Nova Scotian, 151, 152
Lisbon, Portugal, 11, 24
Little King, the, 138, 144, 151, 196, 218, 285, 293, 300; description of, 101-103; fills dory, 184, 223, 224; total catch of, 299
Loch Inver, the, 271
Lousado, the, 261, 267, 277, 284
Luckner, Graf Felix von, 209, 316
Lusiads by Luis Vaz de Camoens, 14
Lutador, the, 252, 282, 287

Maria Carlota, the, 269
Maria da Gloria, the, 16, 200, 245, 269; shelling of, 209-211
Maria das Flores, the, 275
Maria Frederico, the, 261, 268
Maria Preciosa, the, 161, 163, 164
Marine Society, school-ship of, 11
Markland, 39

Marques, Captain Antonio, 72, 181, 286
Marques, David, 73
Martins, Emiliano, 292
Martins, Estrela, 294
Martins, Francisco, First Fisher of Azores, 103, 134, 188, 197, 294; description of, 142, 143; sea experience of, 202, 203; total catch of, 299
Martins, Francisco, sons of, 307
Martins, Jacinto, 138, 188, 197, 203, 234; skill of, 96; sea experience of, 201, 202
Martins, Salvadore, 143, 197, 294, 297
Martins, the Star, 143, 197
Mates, of *Argus,* 130
Matias, Captain, of *Milena,* 72
Matias, João Fernandes, 13, 272
Mauricio, César Eduardo, 13, 63, 90, 129, 142, 159, 161, 272, 305
Medeiros, César de, 203
Medical care, emergency, at sea, 141, 142
Melville Bay, 158
Milena, the, 72, 172, 286
Monica, shipbuilding firm at Gafanha, 26, 31, 121, 213, 320, 321
Morrais, João, 13
Motor-ship: disadvantages of, 115, 208; described, 209; see also Antonio Coutinho, *Capitão Ferreira, Cova da Iria, Elisabeth, Inacio Cunha, João Costa, Santa Maria Madalena, São Ruy,*

*Soto-Maior, Terra Nova,
Vaz* Murtosa, Portugal, 24

Nagsugtok, *see* Norde Stromfjord
Namorado, Tude, 236
National Geographic Society, the, 14
National Society of Codfishing Shipowners, 242
National Syndicate of Captains and Officers, 334
Navegante II, the, 268, 269
Navigation, long-voyage, pioneers of, 37
Nazaré, Portugal: clothes of men from, 23, 24; matriarchal tendencies of, 149
Neptuno, the, 146, 209, 221, 244, 269, 270
Neptuno II, the, 16
Newfoundland: discovery of, 41; taken over by Canada, 76; decline of its fishing, 77 ff.
Newfoundland Fisheries Board, the, 13
Nobre, Manuel de Oliveira, 191
Norde Stromfjord (called Nagsugtok), 273
Normandie, the, 269
Norsel, the, 82
Norsemen, early, 37 ff.
North Sydney, Nova Scotia, 13, 147, 151, 152, 153
Novos Mares, the, 259, 286, 321

O Anjo, Captain Chuva (the Angel), 236, 286, 295
Olhão, Portugal, 191

Oliveira, João de, 67, 134, 143, 170, 203, 224, 294; sea experience of, 197, 198; total catch of, 299
Oliveira, José Luiz Numes de, 13, 272
Oliveira, Sebastian de, 198
Oliveira, Senhora de, 217
Oliveirense, the, 51, 121, 242, 256, 286
Oporto, Portugal, 11, 24, 27
Ornavik, Sweden, 267
Outward Bound Sea School, the, 11

Paços de Brandão, the, ex-*General Rawlinson,* 312
Paião, the, brothers, 200, 208, 219
Paião, Captain Adolfo Simões, Jr., *see* Adolfo, Captain
Paião, Captain Francisco da Silva, *see* Almeida, Captain
Paião, Julio, 219
Paião, Manuel, 219
Pairs, Spanish, 80, 122
Parceria Geral de Pescarias, the, of Lisbon, 12, 314
Parkhurst, Anthony, 76
Pedro de Barcelos, the, 242
Peninsular War, the, 191
Pereira, Duarte Pacheco, 46
Pereira, Dr. Pedro Theotonio, 11, 50
Pereira, Raul, the Footballer, 134, 143, 203; sons of, 307
Phillips, Lt. F. M., 316
Plant, fish-handling, 64
Polar Current, 162
Polaris, the, 163
Ponta Delgada, Azores, 13, 51, 201

INDEX

Ports, fishing, of Portugal, 24
Portugal, language of, 188
Portugal, renascence of, 48
Póvoa de Varzim, Portugal, 27, 31
Primeiro Navigante, the, 270
Prince Regent, the, 191
Programme, fishing, 124

Quintal, 63

Radar, 146, 251
Radio, beam system, 181
Rafael, Manuel dos Santos, 134, 203
Ramalheira, family of, 197, 211, 212, 260
Ramalheira, Captain Anibal, *see* Anibal, Captain
Ramalheira, Elmano, 211
Ramalheira, Captain João Pereira (called Vitorino), 13, 195, 244 ff.
Ramalheira, Manuel, 222
Ramalheira, Captain Silvio, *see* Silvio, Captain
Rancho, 37
Real, João Vaz Corte, 41, 44, 45
Real, Miguel Corte, 46
Rejkjavik, Iceland, 165
Renovation, the, 162
Returns, owner's, 337
Rio Caima, the, 165, 171, 286, 304
Rio Lima, the, 31, 192, 193, 275-277
Rocha, Manuel da Maia, 13, 272
Rodrigues, Antonio, 134, 160, 188, 292; sea experience of, 197-200; opinion of two-men dory, 203; sons of, 295
Rodrigues, Tiago, 58-61

Sagres, the, 26
Sailing, Portuguese, history of, 37 ff.
St. John's, Newfoundland, 13, 68, 73, 74
St. Lawrence, Gulf of, 153
St. Malo, France, 16
St. Michael's, Azores, 51, 52, 202, 305
St. Pierre Bank, 144
Salaries, fishing fleet, 328 ff.
Salazar, Dr., 29, 50; on New State, 301, 302
Salazar, Rebuilder of Portugal by Colonel F. C. C. Egerton, 301 n.
Salgueiro, Don Manuel Trindade, 29
Samos, 128
San Jacinto, the, 222
Santa Isabel, the, 26, 168, 259, 273, 274, 280
Santa Maria Madalena, the, 26, 31
Santa Maria Manuela, the, 31, 68, 148, 166, 280
Santa Quiteria, the, 269, 315
São Jorge, Azores, 305
São Martinho do Porto, Portugal, 24
São Ruy, the, 26, 31
Sargasso Sea, the, 46
Sá Rosa, Father Antonio de, 258
Saudade, 110 n.
School, Fishermen's, 24, 28, 114, 231
Schooners: Portuguese Banks, 20 ff.; Labrador, 78; ad-

INDEX

vantages of, 115, 209; Canadian, 119, 120; Gloucester, 150; *see also Adela Maria, Ana Maria* ex-*Argus, Ana Primeiro* ex-*Erika, Antonio Ribau, Argus, Aviz, Brites, Coimbra, Condestavel, Creoula, Cruz de Malta, Dom Deniz, Gamo, Groenlandia, Hortense, Infante de Sagres, Julia IV, Labrador, Lousado, Maria da Gloria, Maria das Flores, Maria Frederico, Maria Preciosa, Milena, Navegante II, Neptuno, Neptuno II, Novos Mares, Paços de Brandão* ex-*General Rawlinson, Primeiro Navigante, Rio Caima, Rio Lima, San Jacinto, Santa Isabel, Santa Maria Manuela, Santa Quiteria, Senhora da Saude, Silvina, Vencedor, Viriato, W. L. White*
Sea Cloud, the, 211
Sealers, Arctic, 81, 82; lost on ice, 163
Seeadler, the, 209
Seebeck, G., 242
Senhora de Saude, the, 165, 236, 259, 266, 280
Ship, Finnish sailing, 15
Ships, hospital, 243
Signal, recall, 89
Silva, Captain Manoel da, 287
Silva, Manuel Lopes da, 135
Silvina, the, 269
Silvio, Captain, 13, 72, 130, 131, 139, 141, 147, 161, 200; description of, 34; his theory of dory origin, 150;

good fortune of, 177, 277, 278; sea experience of, 205 ff.
Simões, Captain Alfredo, 168, 259, 260, 273, 280
Singing, superstition about, 143
Slater, Alexander, 312
Sociedade Nacional dos Armadores de Bacalhau, 242
Sorrow, Soup of, 97, 228, 291
Soto-Maior, the, 116, 165, 248; described, 148
Sousa, Manuel de, called Vinhas, 134, 138, 151, 188, 197, 225; background of, 200, 201
Sousa, de, the young, 138, 225, 292
Spitzbergen, 123
Stockfish, 38
Store Hellefiske Bank, Greenland, 155, 158, 171, 225, 226, 235, 267, 282, 286
Submarine, German, 210
Sukkertoppen, Greenland, 155, 171, 245, 259
Surgeon, fleet, 141

Tagus, the, 20, 309
Teive, Diogo de, 41, 42, 45, 147
Teive, João de, 41, 147
Telephone, radio, 54, 141
Templeman, Dr. Wilfred, 13
Tenreiro, Commander Henrique dos Santos, 12, 28, 50
Terceira, Azores, 305
Terra Nova, the, 171, 276, 280
Terra Nova, the old, 200
Terror, the, 163

INDEX

Tjalfe, the, 266
Tordesillas, Treaty of, 46
Trawl: defined, 298 n.; new type of, 304
Trawlers, 67, 285; opposition to, 122, 123; French, 123; *see also Alvaro Martins Homem, Aguas Santas, Estavão Gomes, Fernandes Labrador, Genepesca I, Genepesca IV, João Alvares Fagundes, João Corte Real, Pedro de Barcelos*
Tuberculosis, scourge of dorymen, 248

Umanak, Greenland, 225, 282

Vaz, the, 148, 165

Vencedor, the, 221
Vessels, Portuguese sailing, classes of, 320
Viana do Castelo, Portugal, ships from, 11, 26, 31, 68, 276
Vikings, early, 39 ff.
Vinden, James, 12
Vinland, 39
Virgin Rocks, the, 71, 200, 261
Viriato, the, 286

W. L. White, the, 100
Warspite, the, 11
Whalers, Arctic, 81
Whales, menace of to dorymen, 106, 107, 108
Wind system, Atlantic, 46

Zeller, Eduardo van, 12

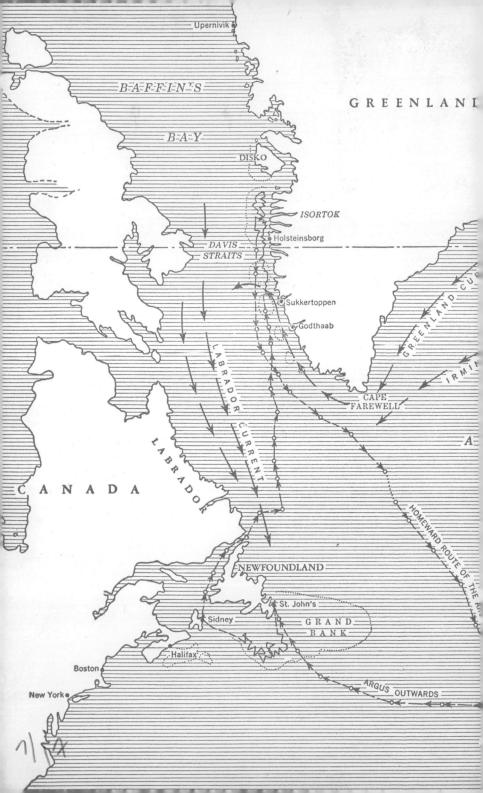